Millimeter-Wave Integrated Circuits

Eoin Carey
Sverre Lidholm

Millimeter-Wave Integrated Circuits

Springer

Library of Congress Cataloging-in-Publication Data

A C.I.P. Catalogue record for this book is available
from the Library of Congress.

ISBN 0-387-23665-1 e-ISBN 0-387-23666-X Printed on acid-free paper.

Printed in the United States of America.

9 8 7 6 5 4 3 2 1 SPIN 11340546

springeronline.com

Dedication

To our wives, Aileen and Phil.

Contents

Preface

The design and implementation of millimetre-wave (mm-wave) integrated circuits using MMIC technology are explored in this work. To make possible the widespread use of MMIC technology for mm-wave applications, high performance MMIC mm-wave transceivers must be shown to be viable for high-volume production. Low noise amplifiers (LNAs), mixers and frequency multipliers, circuit functions critical to the development of a mm-wave transceiver, are studied in this work. The circuit design material presented is augmented by theoretical analyses where possible.

All the designs were fabricated on a commercial 0.25 μm GaAs pHEMT foundry process. The performance realised is compared with simulations for all the fabricated circuits, and where necessary, the causes of significant discrepancies are discussed. All the designs presented are novel to a certain extent. The 40 GHz LNA design approach is particularly novel and outstanding circuit performance has been realised with this design. The conceptual analysis approach presented for HEMT mixer circuits simplifies the understanding of how these circuits work and also aids their design for best performance. A simple but very effective fundamental analysis has been developed for FET frequency multipliers. This approach is successfully applied to balanced multipliers as well. Novel frequency multiplier architectures are proposed that are suitable for the generation of high power levels at very high frequencies. The material presented in this work advances the knowledge base associated with mm-wave integrated circuits. In particular, it demonstrates that high quality circuits can be realised on conventional fabrication processes, thereby suggesting that high-volume mm-wave circuit developments can indeed become a commercial reality.

Eoin Carey (Ó Ciardha)

Sverre Lidholm

Acknowledgments

We would like to acknowledge the support of Farran Technology Ltd. (FTL), and in particular Prof. G. T. Wrixon and Dr. Brendan Lyons for invaluable support. Similarly the support of the Department of Microelectronics, University College Cork, and especially Prof. Peter Kennedy, is greatly appreciated. We would also like to thank M/A-COM Eurotec, and especially Eugene Heaney for supporting this effort.

We would like to acknowledge the assistance of Peter Duffy whose comments were always appreciated and whose inputs should be considered a valuable contribution to this work. We are also pleased to acknowledge the contributions of Sean Cremin (mechanical housing design), Martin Fehilly (precision manufacturing and assembly) and Jason Lynch (test and validation) at FTL.

Thank you to Michael Hackett and Rebecca Olson at Kluwer Academic Publishers for their help and patience in making this work a reality.

Particular recognition is due to our families for their patience and unbounded support during the course of this work being brought to fruition. To Aileen, Cian, Diarmuid, Phil, Heidi and Jack, thank you!

Chapter 1

AN INTRODUCTION TO mm-WAVE INTEGRATED CIRCUITS

1. INTRODUCTION

The primary topic of this work is mm-wave integrated circuits. The main focal points are, firstly, fundamental circuit analyses of some of the building blocks necessary for the widespread employment of MMIC technology for mm-wavelength applications, and, secondly, circuit design methodologies pertinent to these building blocks. The analytical/theoretical treatment is supplemented by specific mm-wave MMIC designs of varied complexity.

2. MOTIVATION FOR mm-WAVES

The term mm-waves is generally used to describe the range of frequencies between about 30 and 300 GHz where the wavelength is of the order of a millimeter. With the ever-increasing demand on wireless spectrum, as evidenced by emerging applications such as third-generation (3G) mobile phones at RF frequencies, and both Multipoint Video Distribution System (MVDS) and Local Multipoint Distribution System (LMDS) in the mm-wave frequency range[1, 2], there is a growing need to exploit higher and higher frequencies. The mm-wave frequency range is very attractive for various applications for a number of reasons. In the first instance, there is inherent in the high frequencies involved, a proportionately large amount of spectrum available. This naturally lends itself to systems demanding wide bandwidths that are simply unavailable at lower frequencies. Secondly, some parts of the mm-wave range have interesting propagation characteristics. For example, the inherent atmospheric attenuation near 60 GHz, due to oxygen absorption, makes this region of the spectrum very useful for short-hop communications links, which are naturally secure and suitable for extensive use with a low likelihood of interference with frequency re-use. Other frequencies have low propagation losses and are suited to longer hop-length links. Additionally, smaller antennas are required at higher frequencies, and this factor, coupled with the good distance resolution capability, has been a key driver in the development of automotive radar sensors at 77 GHz.

3. MOTIVATION FOR MONOLITHIC GaAs INTEGRATED CIRCUITS

The conventional implementation for such mm-wave systems involves extensive use of waveguide technology. Such waveguide components are heavy, bulky and expensive and are not well suited to high volume manufacturing. Therefore, with these emerging volume applications, there is a growing need for monolithic front-end components for the required transceivers. Such monolithic solutions would almost certainly be more easily manufactured in volume, should be physically smaller and lighter and hence more attractive for many size-sensitive applications, and most importantly of all, should reduce the overall cost. Due to the high frequencies involved, much effort has been expended worldwide in developing high frequency semiconductor fabrication capability. Gallium Arsenide (GaAs) has long been recognised for its advantages at high frequencies. GaAs technology is not nearly as mature as that of Silicon (Si), but it does offer benefits in terms of electron velocity that make it extremely attractive for use in mm-wave applications. Initial GaAs circuit developments were based on the Metal Semiconductor Field Effect Transistor (MESFET) transistor. Over time, a number of more sophisticated GaAs based field effect transistor devices, such as the High Electron Mobility Transistor (HEMT) and pseudomorphic-HEMT (pHEMT), have evolved. These have more complex material layering structures than the conventional MESFET, which improve further their high frequency characteristics. GaAs based bipolar transistors, Heterojunction Bipolar Transistors (HBTs), have also been developed which are suitable for high power applications at high frequencies and these make possible levels of performance simply not achievable with conventional silicon bipolar transistors. The availability of such high frequency devices from high frequency fabrication processes facilitates the development of monolithic mm-wave circuit designs for new high volume applications.

4. MOTIVATION FOR IMPROVED FUNDAMENTAL CIRCUIT UNDERSTANDING

GaAs IC fabrication facilities are more expensive to construct than Si facilities, and GaAs material itself is more expensive and more difficult to handle. Moreover, GaAs based ICs are typically developed for specific requirements that cannot be met with a Si solution. The vast majority of worldwide demand for semiconductor products can be more than adequately addressed using Si. As a consequence, GaAs IC fabrication facilities are not as widespread as those based on Si. There are a relatively small number of companies worldwide with their own internal GaAs IC fabrication capability. Some of these companies make their fabrication processes available commercially as a foundry. Due to the high material and wafer fabrication costs, foundry runs can be expensive, in particular for high frequency processes. Moreover, due to the relatively immature science of mm-wave monolithic circuit design, the risk associated with a design can be relatively high, and this level of risk combined with the high cost can result in many mm-wave MMIC developments never getting past the conceptual stage.

In this work, efforts have been made to put mm-wave MMIC design strategies on a more solid foundation. This has been carried out by means of a combination of extensive high frequency design and evaluation, and also a detailed theoretical study of key components of vital interest to the eventual exploitation of MMIC technology for high volume mm-wave applications. In this context, design methodologies for some of the building blocks are proposed.

5. KEY COMPONENTS

Key components of mm-waves transceiver circuits have been studied and evaluated during the course of this work. These circuits include low noise amplifiers (LNAs), a balanced diode mixer suitable for both up and down-conversion, and frequency multipliers. Practical circuit realisations have targeted transceiver applications operating near 40 GHz and 57 GHz.

Specifically, the MMIC circuits, which have been fabricated and tested, are:

- 57 GHz LNA
- 40 GHz LNA
- Balanced Diode Mixer (as both an up and a down-converter)
- Balanced HEMT Mixer as down-converter
- HEMT Frequency Tripler

- HEMT Frequency Doubler
- 57 GHz Transceiver using integrated building blocks

These circuits are represented graphically in Fig. 1-1.

All the above circuits have been designed using the foundry models for the GEC-Marconi Materials Technology (GMMT) H40 GaAs process, and have been fabricated at GMMT's Caswell wafer fabrication facility in the UK using two distinct mask sets. Both wafer runs were implemented as multi-project masks with multiple designs being fabricated on a single wafer. Bookham Technology purchased the Caswell facility from GMMT in February 2002. Refer to[3] for details on the foundry service offered at Caswell.

All the fabricated MMICs were evaluated. Both on-wafer and packaged tests were carried out. The on-wafer evaluations were carried out at the University of Glasgow; the packaged tests were conducted at Farran Technology using specially designed housings. The performance achieved with the various building blocks is very promising. All circuits worked, at least to an extent, at the intended frequencies, and particularly in the cases of the 40 GHz designs, functioned with promising levels of performance. The 57 GHz circuits did not all work as well as predicted, but taking account of the constraints imposed by the GaAs process used and the models available, we consider the achieved performance significant.

6. STRUCTURE OF THIS WORK

This work is structured as follows. In Chapter 2, the characteristics of semiconductor materials suitable for high frequency circuits are described and materials with such properties are outlined. The fabrication technology pertinent to such high frequency material processing is discussed. Diode and transistor devices, based on the material systems discussed in Chapter 2, which are appropriate for high frequency circuit development, are presented in Chapter 3. In particular, the Schottky diode and variants of the MESFET transistor device are detailed.

A primary goal of this work is to improve the fundamental understanding of mm-wave MMIC operation with a view to facilitating the exploitation of MMIC technology for high volume mm-wave applications. In particular, this is driving the demand for fully integrated MMIC transceivers operating in the mm-wave frequency range, and these issues are discussed in more detail in Chapter 4. In particular, the transceiver topology most suitable for implementation in a monolithic design is considered. The concept of yield is introduced and the relationship between circuit yield and circuit complexity is explored.

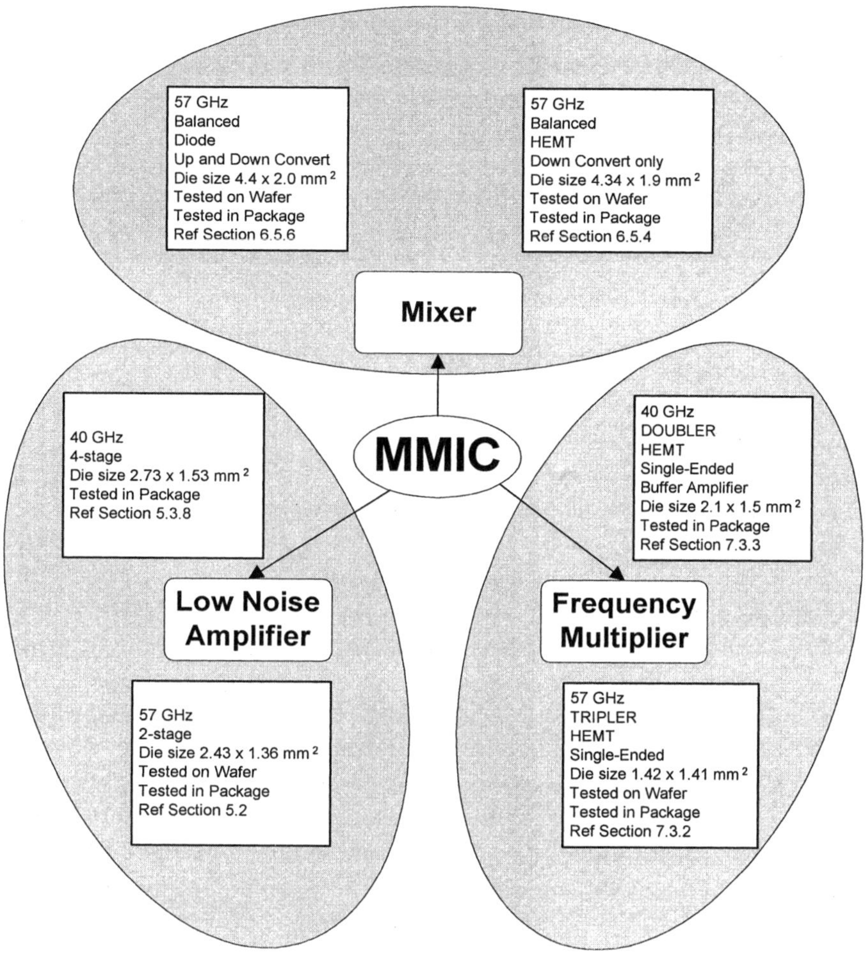

Figure 1-1. Graphical summary of the MMIC designs developed and characterised in this work.

The next three chapters consist of detailed studies of three key building blocks required for mm-wave transceiver developments. In Chapter 5, the Low Noise Amplifier (LNA) is considered. A novel design approach, which is suited to mm-wave circuits, is presented; this is validated by the measured performance associated with a 40 GHz LNA design. This circuit performs very well over a broad band and a good fit is achieved between measurement and simulation. Details of a 57 GHz LNA circuit design are also presented. Monolithic mixers are considered in Chapter 6 and a fundamental understanding of the operation of FET-based mixers presented. Measured results for practical diode and HEMT mixers are also detailed in this chapter. A number of aspects associated with mm-wave frequency multipliers are discussed in Chapter 7; the measured performance achieved with a mm-wave doubler and a frequency tripler are also presented. A novel theoretical treatment of the performance of a generic FET frequency multiplier is outlined, and the measured performance of the practical circuits are compared with that expected in light of the theory presented. The possibilities introduced by balancing are also discussed. A novel balanced multiplier topology, which is suited to a flexible harmonic multiplication factor, is also proposed. The viability of state of the art mm-wave processes for the generation of high power levels at 100 GHz is analysed in the context of a feasibility study and a design topology suitable for a practical realisation is proposed.

In Chapter 8, details are presented of a novel 57 GHz transceiver chip. This is a fully integrated up- and down-converter, and has been implemented by integrating into a single layout a number of building blocks described in Chapters 5 – 7. The performance achieved with this MMIC does not match that expected, and the discrepancy is analysed and discussed. However, the transceiver is functional and this is a highly significant result given the constraints imposed by the capability of the H40 process at 57 GHz.

An in-depth discussion of this work and the contribution it makes to the field of mm-wave MMIC development is presented in Chapter 9. The novel aspects are pointed out, and the key practical and technical achievements listed. The work closes with a discussion of likely future trends in high frequency MMICs, both in terms of circuit design and also in terms of process and device developments.

Chapter 2

HIGH FREQUENCY MATERIALS AND TECHNOLOGY

1. INTRODUCTION

Silicon (Si) is undoubtedly the workhorse semiconductor material of the electronic age. It has many characteristics, both chemical and electrical, which have contributed to its unrivalled status – see for example Sze[4]. Both bipolar and metal oxide semiconductor (MOS) type devices, fabricated on Si substrates, are widely used for a large number of applications[4]. As we will see in Section 3.2 of this chapter however, Si based devices are constrained in terms of their performance capability at high frequencies, in particular in the microwave range and above. In this chapter, the characteristics of semiconductor materials more suited to high frequency devices are discussed. These characteristics are firstly presented in a generic sense, essentially defining the 'ideal' high frequency semiconductor material. Subsequently, the characteristics of some of the high frequency materials in use today are outlined. These materials tend to be III-V compound semiconductors, and the reasons for this will be discussed. Some of the key wafer fabrication techniques involved in the processing of high frequency materials, and particularly pertinent to their high frequency characteristics, are described. Finally, some of the main aspects associated with the realisation of effective mm-wave MMIC layouts are discussed. These are generic considerations which need to be taken into account throughout the circuit design effort in order to ensure that the resulting layouts are not needlessly constrained from a high frequency performance perspective. These considerations include a description of circuit layout techniques to facilitate circuit testing and subsequent wafer sawing or dicing.

2. ELECTRICAL CHARACTERISTICS OF IDEAL HIGH-FREQUENCY SEMICONDUCTOR MATERIAL

There are a number of electrical characteristics that render a semiconductor material suitable for high frequency applications development. A device with high carrier mobility, μ, is desirable. Such a material will be responsive to rapid changes in an applied electric field. A quantum mechanical analysis of semiconductor material band structures shows that the mobility of a material is dependent on the curvature of the band valleys and peaks[5]. As a consequence, in order for high mobility to be achieved, a semiconductor should have sharply curved bands. It is the case with semiconductor materials that the electron mobility exceeds that of the holes and this is particularly true of high frequency semiconductor materials. It is therefore not surprising that most high frequency devices use electrons – see Chapter 3.

It is known that there is a definite correlation between the mobility of carriers in a semiconductor and the material's energy band-gap. In fact, it is found that higher mobilities are generally associated with smaller band-gaps. As a consequence, one could reasonably expect that a small band-gap would be an essential feature of a high-speed semiconductor. However, a small band-gap would also lead to a relatively high leakage current due to a large number of carriers having sufficient thermal energy to traverse the energy gap. Accordingly, an ideal semiconductor has as small a band-gap as is practical such that leakage current does not become a concern.

It is intuitively clear that in order that a material be suitable for high frequency circuit developments, high carrier velocities must be possible. Such high velocities result in short transit times for carriers to traverse the device geometry, thereby making the device responsive to high frequency excitations. There is a limitation to the carrier velocity which can be realised in a given semiconductor material. From a purely relativistic standpoint, the carrier velocity cannot exceed the speed of light, c. In real materials, scattering events ensure that carrier saturation velocities are in fact much less than c. These scattering events can be associated with collisions with the lattice, including dopant sites. Scattering events lead to sudden changes in the carrier velocity and energy, and the scattering path describes how the carrier shifts from one high-energy band to another (low-energy) band. When an increasing field is applied to a material, the field accelerates the carriers. As the field increases, different scattering mechanisms become important. The result is that the rate of carrier acceleration falls as additional scattering effects come into play. At medium fields, the carriers may have sufficient energy to excite acoustic phonon vibrations in the material lattice.

This effect is especially important in indirect band-gap materials where the large number of conduction band minima increases the likelihood of such an event occurring. Higher fields can lead to the excitation of optical phonons. Further field increases lead to energy being transferred directly to the lattice (heating) and the carrier velocity saturates. In order that the onset of saturation occurs at high fields, and hence high carrier velocities can be achieved, a direct band-gap material is necessary.

An important property of high quality semiconductor devices is carrier confinement or localisation. This requires that the physical space in which carriers traverse the device is well controlled and understood. In early devices, this confinement was achieved by the use of p-n junctions. These devices are constrained in their high frequency capability due to hole mobility limitations. For many of the widely used high frequency devices, a Schottky junction provides the confinement. Band structure plays a major role, particularly in the case of high frequency materials. This is brought about by the use of hetero-structures. A hetero-structure is a combination of at least two semiconductor materials with different band-gaps. The differing band-gaps introduce some interesting possibilities, as will be explained in Section 2.4 of Chapter 3 for the high electron mobility transistor (HEMT) device. In the HEMT, the localisation is realised as a two-dimensional electron gas in an undoped GaAs layer. This makes possible the major advantage of having the channel in undoped GaAs, which has a much higher mobility than doped GaAs. In order that hetero-structures can be fabricated between two semiconductor materials, they must have similar lattice constants (i.e. they are approximately lattice-matched). This is required such that atomic bonding can be continued, without interruption, across the interface. Thus, an ideal semiconductor should be lattice matched to at least one other material.

In conclusion, a high-speed semiconductor should consist of a direct, narrow band-gap material (but not too narrow) lattice-matched to at least one other such material. Compound semiconductors, such as GaAs, come closest to matching these ideal requirements.

3. ELECTRICAL CHARACTERISTICS OF REAL HIGH FREQUENCY MATERIALS

In this section, a brief outline of the major compound semiconductors is presented. Firstly, the electrical properties of GaAs are discussed, in particular in the context of a comparison with silicon. Subsequently, properties of some of the other compound semiconductors used at high frequencies are considered, and the current trends in high frequency device developments are discussed.

3.1 Gallium Arsenide (GaAs)

Gallium Arsenide (GaAs) has gained wide acceptance as the semiconductor material of choice for high frequency applications, in particular in the microwave and mm-wave ranges. Since the early GaAs development activities, a number of device structures have evolved which exploit its excellent high frequency characteristics. The GaAs Schottky diode remains a very commonly used device, particularly at high mm-wave and even sub mm-wave frequencies, where active devices with gain cannot yet be realised. The non-linear diode characteristic makes this device a very useful candidate for high frequency mixers, detectors and frequency multipliers, and thus the Schottky diode plays a key role in very high frequency receivers[6]. At somewhat lower frequencies (RF and microwave), the GaAs MESFET is the conventional active transistor device. Due to certain limitations associated with the MESFET, other GaAs based FET-type transistors have evolved, including the HEMT, the p-HEMT, the lattice-matched HEMT (on InP substrate), and the metamorphic HEMT. The bipolar-type HBT transistor has also emerged as a device with great high frequency potential. These various high frequency GaAs-based circuit devices will be discussed in detail in Chapter 3.

3.2 GaAs / Si Comparison

The dominance of GaAs as the primary semiconductor material for high frequency applications is due to its excellent physical parameters, some of which are compared with those of Si in Tables 2-1 and 2-2[7]. It should be noted that these comparisons are valid for specific doping densities in the two materials – 10^{16} cm^{-3} and 10^{17}cm^{-3} in Tables 2-1 and 2-2 respectively.

Table 2-1. GaAs/Si parameter comparison ($N_d = 10^{16}$ cm^{-3}, 300K)

PARAMETER	GaAs	Si
Saturated Drift Velocity (cm/s)	1 –2 x 10^7	0.7 x 10^7
Electron Mobility (cm^2/V/s)	5000	1300
Hole Mobility (cm^2/V/s)	330	430
Band-gap (eV)	1.42	1.12

The tabulated mobilities correspond to *low-field* operation. It can be seen that both the mobilities and the saturated drift velocity parameters are functions of the doping level. The greater the doping concentration, the lower the mobility, due to an increased likelihood of collisions with the dopants. The velocity versus field curves associated with Table 2-2 are shown in Fig. 2-1. This Figure also includes the corresponding curve for InP.

Table 2-2. GaAs/Si parameter comparison ($N_d = 10^{17}$ cm^{-3}, 300K)

PARAMETER	GaAs	Si
Saturated Drift Velocity (cm/s)	8×10^6	6.5×10^6
Electron Mobility ($cm^2/V/s$)	3500	800
Hole Mobility ($cm^2/V/s$)	250	300

In fact, the mobilities of *undoped* GaAs are 8500 $cm^2V^{-1}s^{-1}$ (electrons) and 400 $cm^2V^{-1}s^{-1}$ (holes)[8]. This is of great significance in some hetero-junction based devices to be discussed later in Chapter 3. Additional mobility values as a function of doping density are shown in Table 2-3.

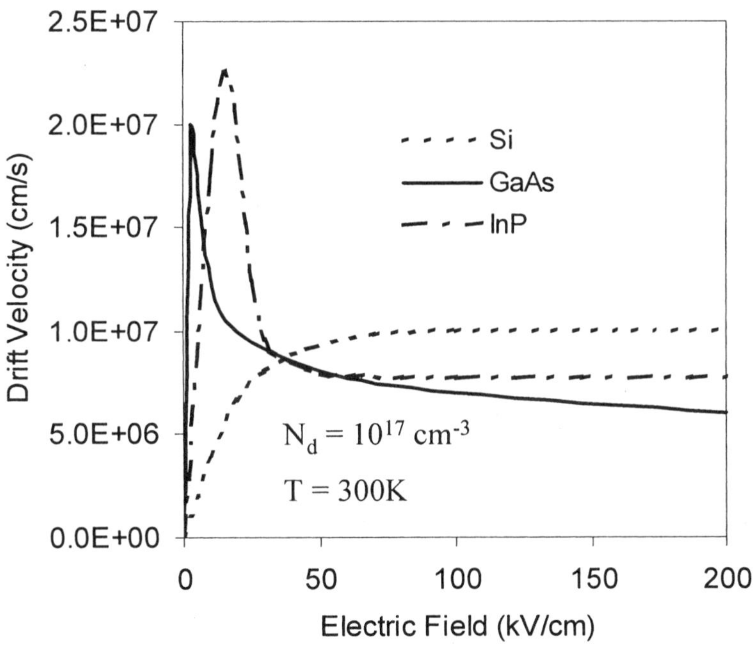

Figure 2-1. Electron drift velocity versus electric field ε.

It should be borne in mind that for many applications, the saturation velocity is of primary interest. However, for very high frequency applications, the much higher carrier velocities in compound semiconductors at lower fields are of interest due to the overshoot effects observed in devices of very small size.

GaAs has a higher intrinsic power delivery capability than Si. This can be seen directly from the following expression for the power-frequency-squared limit[9],

$$Pf^2 \approx \left(\frac{E_c v_s}{2\pi}\right)\frac{1}{X_c}$$

where E_c is the effective electric field before avalanche breakdown, v_s is the electron drift velocity, and X_c is known as the *device impedance level.* Since E_c and v_s are higher for GaAs than for Si, it follows that the power-frequency-squared limit is also higher for GaAs.

Table 2-3. GaAs electron mobility as a function of doping density

Doping Density cm^{-3}	Electron Mobility (cm^2/V/s)	Hole Mobility (cm^2/V/s)
Undoped	8500	400
10^{16}	5000	330
10^{17}	3500	250
10^{18}	2000	-
10^{19}	300	80
3×10^{19}	-	40

Applying the above relationship for GaAs and Si material, and using typical values for X_c (based on the material and the device structure), the following estimated theoretical limits have been reported[9]

$$\text{GaAs} \rightarrow Pf^2 \approx 5 \times 10^{21}\,\text{Ws}^{-2}$$

$$\text{Si} \rightarrow Pf^2 \approx 5 \times 10^{20}\,\text{Ws}^{-2}$$

This suggests that at a given frequency, for similar device geometrical structures and sizes, a GaAs based active device is capable of delivering about 10 times as much power as its Si equivalent. Alternatively, the GaAs device is capable of driving a given power level at a maximum frequency which is more than 3 times the maximum frequency for the equivalent Si device. This explains why GaAs is the preferred material at mm-wave frequencies.

GaAs is also capable of providing more gain than Si. This is mainly due to its higher electron mobility, which implies that, for a given electric field, a greater electron velocity will be achieved in GaAs, in particular due to overshoot effects in small devices at high frequencies. This can be interpreted in terms of a greater output load current flowing for a given applied input voltage, which is associated with a higher gain.

An additional characteristic of GaAs is that the GaAs substrate material is a very good electrical insulator. This can be a very important factor in the development of high frequency monolithic circuits, which are typically microstrip-based. Si substrates are much poorer insulators and as a consequence, passive structures on Si tend to be much lossier than their GaAs equivalents.

GaAs based amplifiers are capable of providing a lower noise figure than Si. This is due to a number of factors. Primarily, the greater mobility means that the random noise events (e.g. collisions) are less significant relative to the drift currents. Moreover, a FET type device (most GaAs high frequency circuits utilise FET type devices) contains fewer sources of noise (e.g. no shot noise). As outlined by Pavlidis[10], the presence of capacitive coupling between the gate and the channel in a MESFET type structure results in the overall noise being determined by subtracting part of the gate noise from the drain noise. This is a unique property of FETs and can lead to very low noise performance.

Another important distinction between GaAs and Si is that GaAs is a *direct band-gap* material[4]. This means that the minimum energy separation between the conduction and valence bands occurs at the same momentum. Si, on the other hand, is an *indirect band-gap* semiconductor, which implies that its conduction band minimum is separated in momentum from the valence band maximum[7]. This direct band-gap property is typical of many compound semiconductors and their alloys, and is critical to the opto-electronic operation of these materials. Electron - photon interactions are much more efficient in the direct band-gap materials as they do not require an associated phonon scattering event. The direct band-gap also leads to some desirable consequences for electron transport, and in particular is consistent with the high electron velocities achievable in GaAs. A visual representation of the direct/indirect band-gap properties of GaAs and Si is presented in Fig. 2-2.

For completeness, it is appropriate to mention that the cause of the unusual shape to the velocity – electric field curve for GaAs is that a secondary conduction band minimum (or valley) exists which is offset in momentum from the valence band peak. This secondary minimum is at a higher energy than the direct primary valley. The mobility associated with electrons in this secondary valley (refer to Fig. 2-2) is lower than in the main valley because of the secondary valley's much lower curvature characteristic. Thus, as the electric field is increased, and some of the electrons achieve sufficient energy to make the transition to the secondary valley, their mobility (and hence velocity) falls. As the field increases further, more of the electrons can make the transition, and the average velocity continues to tail off towards a lower asymptotic limit. This falling velocity for increasing field (above the critical field) is a phenomenon peculiar to some compound semiconductors. It can be modelled as a negative differential resistance which makes bulk material suitable for the generation of high frequency oscillations under certain conditions (such an oscillator is known as a Transferred Electron Device; a well-known example is the Gunn oscillator).

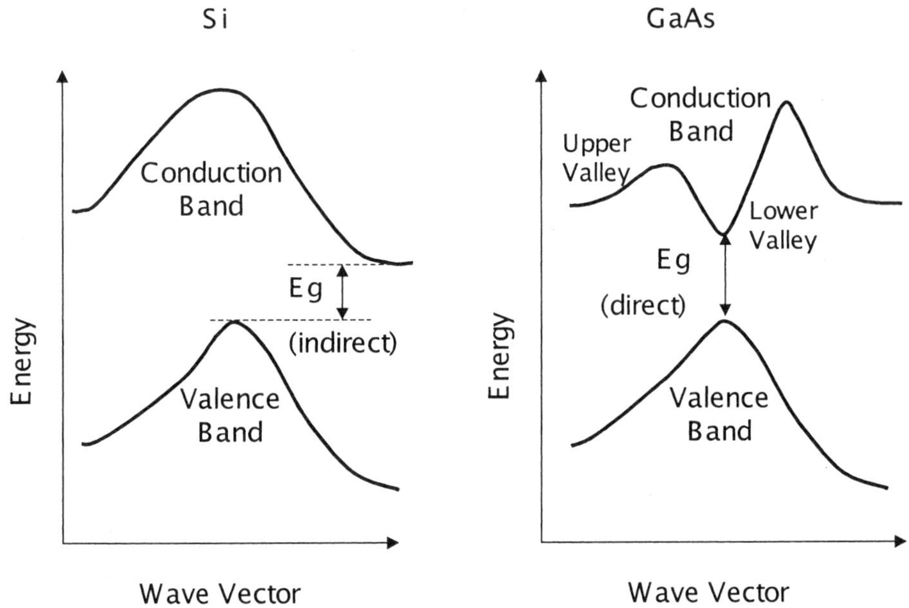

Figure 2-2. Direct and indirect band-gap characteristics of Si and GaAs.

Of course, GaAs does have disadvantages. Si based fabrication processes, being substantially more mature, are cheaper to develop, install and maintain, and are better characterised and understood. Reliability and yield optimisations have been performed much more extensively on Si processes. Silicon material is intrinsically more stable, and has the major advantage of having an excellent native oxide.

Silicon-Germanium (SiGe) hetero-structures with Si are a topic of significant research at present. They offer the potential for many of the benefits associated with hetero-structures in general while, at the same time, they are largely compatible with standard Si IC processing techniques. SiGe transistors with promising high frequency capability have been reported recently[11].

3.3 InP

It has been recognised for a long time that Indium Phosphide (InP) material is blessed with a number of excellent high frequency characteristics. In particular, the peak electron velocity associated with bulk InP is significantly higher than the corresponding values for either GaAs or Si, see Fig. 2-1. As a consequence, for a MESFET type device with a given gate length, the transit time for electrons under the gate is potentially lower in InP. It immediately follows that the transition frequency, f_t, which is defined and discussed in Chapter 3, Section 2.6, is higher (for a gate length of 1 μm, the f_t of InP is 48% higher than that of GaAs), and one might reasonably expect that the InP-based device should be capable of performing at higher frequencies. In much the same way that the transit time for InP is smaller than for GaAs, it is found that InP also has a higher value for transconductance, g_m, for the same bias current[12].

It is indeed true that the use of InP material is beneficial in some application areas. For example, InP bulk material is commonly used in transferred electron devices like Gunn oscillators. When used in this way, the domains generated can traverse the device more rapidly and hence more domains are generated per unit time than for a similar GaAs structure. It then follows that higher frequency oscillators can be developed.

However, in the case of integrated circuit technology, the expected performance advantages of InP over GaAs have not manifested themselves due to a number of factors, including non-optimum material characteristics, buffer layer and substrate quality problems, and technical issues associated with the low barrier characteristics of InP Schottky gates.

The InP gate electrode has a low barrier height. This low barrier leads to an increased leakage current due to thermally excited electrons. The reverse I_{gd} for an InP based MESFET type device is 1000 times larger than that for an equivalent GaAs device[12]. It should be noted that the breakdown voltage of InP is somewhat greater than that associated with GaAs[13].

Another consequence of the smaller energy band-gap is the fact that InP is a poorer insulator than GaAs. Typically, the resistivity of InP is 10,000 times less than that of GaAs[12]. This reduced resistivity has a very significant effect. The current through the substrate at large V_{ds} is greater, and the output conductance is increased. As discussed elsewhere, an ideal device has zero output conductance.

The gate-drain capacitance of an InP device is much larger than that of its GaAs equivalent. It has been explained[12] that this is due to the fact that InP requires a higher field for the onset of velocity peaking and the associated domain formation. As a result, for the bias levels generally used, the Gunn domain formation is weaker (in fact, if the drain bias were increased sufficiently to bring about velocity peaking, the gate drain breakdown voltage would probably be exceeded). It follows that the depletion layer capacitance is not as effectively decoupled at the drain, and hence C_{gd} for InP can be as much as five times its GaAs value.

Much work has been carried out looking at alternatives to doping InP with chromium (Cr) which is generally used to realise a semi-insulating substrate (see Section 5.1 for more information in the context of GaAs). This includes studies of Fe doping which results in increased InP resistivity values[12]. Similar research carried out looking into possible solutions for some of the other InP issues described above has also been reported. Despite much effort, however, it has been generally concluded that InP MESFET type devices do not offer any significant advantages over GaAs equivalents. Instead, most of the research over recent years has concentrated on the technology required to produce FET devices based on the superior properties of some of the ternary or quaternary compounds[14]. Ironically, to date it has been found that the optimum solution, involves a material structure which is best suited to growth on InP substrate material, so that InP has a very important role to play after all (though not as the active channel material) in some current state of the art high frequency devices – see Chapter 3, Section 2.4.3.

3.4 Other III-V Compound Semiconductors

Due to certain limitations associated with GaAs (and particularly InP), considerable work has been carried out in recent years assessing various compound semiconductor materials. A number of material systems have been widely studied[15], and in this section, some key characteristics of a few of the more important materials will be detailed.

A fundamental characteristic of any semiconductor material is its lattice constant. The lattice-constant of a semiconductor is a measure of the length over which a unit-cell of the semiconductor repeats itself in the crystal. The viability of growing different materials on each other with good crystalline quality is dependent on how well their lattice constants match. When two materials have the same lattice constant and one is grown on the other, the resulting hetero-structure is termed lattice matched. If the lattice constants are not the same, some strain will result at the interface. Provided the lattice constants are within about 8% of each other, a high quality crystal growth can still be achieved with careful processing. Such a slightly mismatched material system is termed pseudomorphic. A useful comparison of lattice constants (and band-gaps) of some III-V materials is presented in Fig. 2-3.

It can be identified immediately from Fig. 2-3 that GaAs and AlGaAs are very well lattice matched. Similarly, InP is well matched with specific mole fractions of GaInAs and AlInAs. The significance of these material mixes will become clearer shortly.

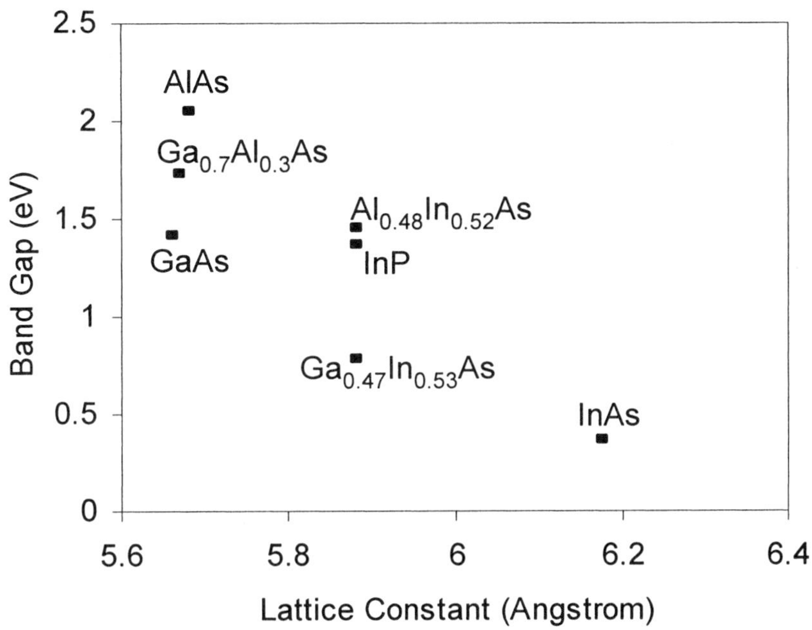

Figure 2-3. Band-gaps and lattice constants for compound semiconductors.

3.5 InGaAs

It is well known that $In_{0.53}Ga_{0.47}As$ exhibits some excellent high frequency characteristics[14]. For example, this material has a room temperature electron mobility approximately 55% higher than GaAs for a doping concentration of 1 x 10^{16} cm^{-3}. This material is lattice matched to InP (Fig. 2-3); hence the use of InP as a semi-insulating substrate is becoming more common for state of the art mm-wave devices.

However, it is found that a Schottky contact of Al on $In_{0.53}Ga_{0.47}As$ has a barrier height of only 0.3V (0.7 - 0.8 V for GaAs). This would lead inevitably to increased tunnelling currents and higher leakage. As a consequence, the barrier height is generally increased by means of a thin intervening layer. This means that InGaAs is not suitable for use as the channel in a conventional MESFET structure, where it would be in direct contact with the gate metal. In fact, it is much more suited to hetero-junction style FETs, as will be discussed in Chapter 3.

4. III-V COMPOUND SEMICONDUCTOR FABRICATION TECHNIQUES

There exists abundant literature on semiconductor fabrication process techniques and technology[16]. The bulk of the published information relates to silicon processing, but a significant volume of work also addresses the specific issues associated with compound semiconductor, and GaAs in particular, processing[8, 14]. In this section, a brief summary of the key fabrication steps relevant to GaAs circuit fabrication will be presented. The particular demands placed on the fabrication technology by the specific requirements of high frequency devices will be discussed in more detail. In this context, the current trends and likely future developments in fabrication technology will also be mentioned.

We have explained previously that GaAs substrates can be manufactured with very high resistivities. These can be used to fabricate both active and passive devices with very low parasitic capacitances to ground, thereby extending the frequency range of operation. The high frequency circuits considered throughout this work are typically fabricated on a semi-insulating GaAs substrate. An n-type epi-layer is formed on top of the semi-insulating substrate, with the active components being manufactured in this epi-layer. This layer can be created either by ion implantation (into the substrate itself), or by epitaxial growth on top of the substrate. Ion implantation is emerging as the preferred option for volume applications. Epitaxial growth can yield superior profile control and is used in applications where the required profile cannot be realised using ion implantation; for example, molecular beam epitaxy (MBE) can be used for the accurate definition of the very thin layers required for some of the δ-doped hetero-structures discussed in Chapter 3. As is the case with all semiconductor devices, the performance characteristics of fabricated active devices are strongly dependent on the doping profile.

A typical high frequency FET contains ohmic and Schottky contacts. For analogue MMIC processes, the ohmic contact is generally formed initially, followed by a gate etch and Schottky contact fabrication. The ohmic source and drain are usually formed using standard optical lithography techniques. The gate metal used to be exposed similarly; however, with the demand for higher frequencies of circuit operation, shorter gate lengths are required and electron beam lithography is now commonly used for this part of the process due to the improved definition it offers. The finished circuit is usually passivated. This provides two main benefits. Firstly, the surface of the GaAs is stabilised. Secondly, the passivation provides some level of protection in the harsh environments associated with handling of the wafer after processing.

The rapidly growing use of GaAs material in both analogue and digital electronic circuits is placing an increasing emphasis on the requirement for large area and high quality wafers with good temperature stability and good uniformity. Due to the typically high capital costs associated with GaAs processing equipment, circuit yield is a critical concern, and this demands a low density of dislocations or imperfections in the crystal structure.

5. GaAs FABRICATION TECHNOLOGY

5.1 Crystal Growth

The material properties of Gallium Arsenide (GaAs) are covered in detail in a number of works, including[17]. Gallium (Ga) is a rare element, and is toxic. Standard purification processes make it possible to derive Ga as pure as 99.99999%. In its liquid form, Ga reacts with quartz at high temperatures, thereby yielding impurities in GaAs grown in quartz containers.

Arsenic (As) is primarily obtained from sulphur ores. It can be obtained in three forms; metallic crystalline (most stable), yellow crystalline and amorphous (an amorphous state essentially means that the element atoms are arranged randomly, as distinct from a regular lattice type arrangement found in the crystalline state). It is more difficult to purify than Ga and is highly toxic. Its vapours are chemically very active.

GaAs growth is complicated by a number of factors:

- It decomposes or dissociates when heated above about 600°C which limits the temperature to which GaAs material can be heated during processing
- Chemical interactions with container materials
- Expansion during solidification can cause stresses

The most common method for growing GaAs is to 'pull' the crystal from the melt (Czochralski method). Mechanisms are required to bear down or press on the melt, and thus prevent dissociation, to enable working temperatures in excess of 600°C. The liquid encapsulation technique (LEC) is now the standard method of achieving this. A layer of boric oxide, a thick glassy substance, covers the melt and prevents decomposition.

Unintentionally doped bulk GaAs grown at 1238°C (melting point) is normally n-type due to Si donors from the quartz container or some original chemical. By adding chromium (Cr) to the melt, chromium deep acceptor traps can be obtained with a density up to the solubility limit. These intentional deep acceptor sites trap the unintentional donors. Consequently, a resistivity as high as 10^8 Ω-cm can be achieved. The high concentration of ionised impurities leads to a lower mobility. The high dislocation and trap density can lead to traps diffusing into the active layer grown on the substrate. A substrate bake at 750°C for 20 hours sharply reduces the diffusion of the traps into the active layer. However, this bake may cause the surface of the substrate to become p-type (possibly due to As being lost in the bake, and Si movement from Ga to As sites). Replacement of Si with intentional tellurium donors helps to prevent this problem. The resulting GaAs substrate exhibits excellent insulation properties.

5.2 Epitaxy

Over the years, a number of distinct epitaxial growth processes have evolved. Most of these were initially used in the silicon semiconductor processes and were subsequently applied to GaAs material. Due to the entirely different nature of the GaAs material, the processes have necessarily been re-engineered and re-optimised for this application.

5.2.1 Liquid Epitaxy

This involves liquid Ga being saturated with As. The saturation temperature of this mix is in the region of 850K. On cooling, GaAs is precipitated due to the reduced saturation level. By suitably arranging an existing GaAs substrate in the growth chamber, the precipitated material forms on the substrate. This heating/cooling cycle may be repeated to grow a series of layers on the substrate (suitable for multi-layer heterojunction applications).

A significant advantage of the liquid phase epitaxy (LPE) process is that the grown material may have a lower density of impurities than the components provided, as any impurities tend to remain in the liquid phase.

5.2.2 Vapour Epitaxy

In this case, the raw materials for the epi-growth are supplied in a gaseous rather than liquid phase (these gases can be derived from either the solid or liquid phase by heating). The gases flow over an existing substrate at a lower temperature, where deposition occurs.

5.2.3 Metal-Organic Chemical Vapour Deposition (MOCVD)

This is a variation of the standard vapour expitaxy where the Ga source is either a trimethylgallium or triethylgallium organic material. Such a process does not require a hot reactor - only the substrate requires heating for MOCVD. The growth rate for this process is slower than for either regular vapour epitaxy or liquid epitaxy.

Due to its slow growth rate, MOCVD has the distinct advantage of being ideally suited to the growth of hetero-junctions with good control over the thickness of thin layers. Such structures play a key role in a number of applications, including high frequency HEMTs and a broad range of opto-electronic devices.

5.2.4 Molecular Beam Epitaxy (MBE)

This is a newer addition to the epitaxial growth family of processes. This technique involves the controlled evaporation (or co-evaporation) of one of more thermal sources to deposit films under high vacuum conditions. The source materials are provided by means of atomic or molecular beams emitted by effusion cells. These cells are essentially heated containers with apertures facing the substrate. A number of shutters are located in the apparatus; one main shutter near the substrate, and one controlling the output beams from each effusion cell.

In MBE, the substrate temperature may be kept relatively low (500°C - 650°C) leading to low growth rates, of the order of 1 $Ås^{-1}$. The beam flux or flow rate can be changed very rapidly by means of the shutters, making it feasible to modify the composition or doping of the grown structures literally within one atomic distance. By rotating the substrate during growth, a non-uniformity of better than 2% can be achieved in film growth for a 2" wafer. Nominally undoped GaAs MBE layers are p-type[18].

The electron concentration of intentionally n-doped GaAs (Si doping yields the highest liquid nitrogen cooled mobility, and is hence the most common dopant) produced by MBE can be more than $5\times10^{18} cm^{-3}$. This is considerably higher than what can be achieved in layers grown by vapour epitaxy where saturation-induced precipitation effects limit the doping density possible. Such highly doped layers are suitable for the production of non-alloyed ohmic contacts (critical for low resistance contacts to the semiconductor devices).

MBE also has the advantage of smoothening the surface of the GaAs layer during growth. This makes this technique very attractive for the growth of hetero-junctions, superlattices and other multi-layered structures, where the planar characteristics of each layer depend strongly on the planar characteristics of the layer underneath.

5.2.5 Epitaxial Growth - mm-wave Implications

As will be described in Chapter 3, hetero-junction based transistors are a key requirement for many high frequency MMIC developments. For many years, MBE was considered the only viable option for the fabrication of hetero-structures of sufficient quality for use at mm-wave frequencies. However, in recent years, MOCVD techniques have been refined and this technology is now capable of delivering high quality hetero-structures to rival the MBE approach. The capital outlay associated with a MOCVD reactor is significantly less than for the corresponding MBE equipment, and it is likely that MOCVD will continue to make inroads into what used to be considered exclusively the domain of MBE processing.

5.3 Ion Implantation

This technique has developed into one of the most important for fabricating GaAs microwave devices and integrated circuits. It can be used to achieve either doping or isolation. The doping density and distribution are controlled by the ion flux and velocity. The defects created during the bombardment are annealed at about 800°C. Ion implantation is especially important for compound semiconductors like GaAs where diffusion doping is more difficult than for Si. In this context, ion implantation is a key technology driver for high frequency device development.

The advantages of ion implantation are that independent control is achieved over the doping level and profile, good reproducibility is possible, and selective doping of specific areas on the surface can be achieved using conventional masking techniques. Multiple implants can be employed to create complex doping profiles, with good lateral resolution (which may not be possible with epitaxial techniques). The distribution of dopants after the implantation is approximately Gaussian in shape, and is characterised by the average projected range R and its associated standard deviation ΔR, which is a measure of the spread in R.

The standard ion implantation model predicts a Gaussian shape for amorphous targets, where the deflection of an ion is purely random. In crystalline targets, the distribution depends on the crystallographic orientation. Channelling can occur if the direction of the ion travel is aligned with the crystal axis - essentially the ions can 'squeeze' between rows of atoms in the crystal and thus travel much further into the crystal. This can lead to non-Gaussian profiles, second peaks, etc., which make the prediction of the resulting doping profile difficult. This problem can be avoided by deliberate misalignment of the beam with respect to the crystal axis.

The doping efficiency, η, depends on the dose. A higher η is found at low dose due to less material damage.

Implanted ions displace semiconductor atoms from their sites. If the energy is sufficiently high, areas of amorphisation, where the crystal structure is damaged, are created. At a large implant dose, the surface layer may become amorphous. In GaAs, amorphous layer formation is suppressed if the ions are implanted at a temperature in excess of 150°C. Alternatively, the material can be annealed at about 800°C afterwards.

Arsenic is lost from the surface of GaAs at temperatures above about 600°C. Plasma deposited Si_3N_4 can be used as an encapsulant, which allows the annealing of GaAs up to 950°C. At up to 750°C, AlN has been found to be an even better encapsulant because of a smaller difference with GaAs in thermal expansion coefficients.

Another technique involves annealing in As vapour overpressure – this is an example of capless annealing. Rapid thermal annealing (RTA), involving the rapid cycling of banks of infrared lamps[14] is another capless technique.

The diffusion of dopants during annealing modifies the Gaussian distribution of the impurities. Many encapsulants act as barriers for out-diffusion, leading to an increase in doping concentration near the surface compared to the original near-Gaussian profile.

The quality of the original substrate is critical, particularly at the low doses used for the active layer of FETs (total dose $< 10^{12}\,cm^{-2}$). The possible complications include the formation of a surface conducting layer in the non-implanted regions of the semi-insulating substrate during anneal (known as surface conversion). Also, the ion-implanted profile may vary across the wafer and extended tails in the profile may appear. An indication of poor substrate quality may be an apparent activation energy of implanted species greater than 100%. Careful substrate qualification tests help to eliminate these problems, which may be related to the diffusion of Cr in Cr-doped semi-insulating substrates.

Semi-insulating GaAs may be produced by ion implantation in the form of ion induced damage. For example, proton bombardment is used to isolate IMPATT structures, lasers can be used to form simple optical waveguide structures and make GaAs ICs using epitaxial material, and oxygen ions can be used to produce semi-insulating GaAs layers.

However, for mm-wave applications, critical GaAs doping profiles are primarily realised using MBE or MOCVD techniques. Nevertheless, ion implantation is still used extensively for isolation and for the fabrication of ohmic contacts.

5.4 GaAs Dopants

Whether epitaxial growth or ion implantation techniques are employed, the commonly used n- and p- type dopants for GaAs are outlined below.

5.4.1 N-Type Dopants

The n-type dopants used for GaAs are:

- Se, Te, S from group VI
- or Si, Sn from group IV

Group VI atoms fill As.

Group IV atoms fill Ga sites. In the case of ion implantation, the doping efficiency for Sn is a strong function of the temperature, whereas that for Si is almost independent of temperature.

5.4.2 P-Type Dopants

The elements used are: Zn, Be, Cd, Mg, C.

For low doses ($< 10^{14} cm^{-2}$), the doping efficiency is about 100%. For higher doses, η falls due to dopant solubility limit being reached. Note that Cr is only suitable as a deep acceptor trap to neutralise unintentional n-type doping.

5.5 Schottky and Ohmic Contacts

A Schottky contact refers to the junction between a moderately doped semiconductor region and a metal. When the semiconductor is very heavily doped, the electrical characteristics of the junction change, and an ohmic contact results. In the case of high frequency devices, Schottky contacts are used for diodes and the gate terminals of FETs, while ohmic contacts play a key role in the source/drain terminals of FETs, and in all three terminals of a HBT.

5.5.1 Practical Schottky Contacts

The semiconductor material is etched just prior to metal deposition (normally by sputtering). After this etch, it is inevitable that an undesirable thin interfacial oxide layer is formed. The resulting surface states mainly determine the barrier height. The choice of metal is mainly driven by such factors as ease of deposition, reliability, resistance, etc. Aluminium (Al) is widely used on GaAs. Tungsten (W) contacts are used when high processing temperatures are required (for example, annealing after ion implantation with the contact in place). However, it yields a higher contact resistance than Al. TiW/Au is also popular. TiW alloys have good adhesion properties, are corrosion resistant, and provide a diffusion barrier, while Au decreases the contact resistance[18]. This resistance is particularly important for high frequency devices, which tend to have small gate length dimensions. Additionally, at high frequencies, the skin effect increases the effective gate metal resistance, thereby further emphasising the demand for minimal resistivity. Furthermore, for low noise amplifiers at high frequencies, where the intrinsic noise characteristics of the transistors naturally degrade, the thermal noise associated with the parasitic gate resistance in FETs must be minimised in order to yield an acceptable noise figure for the extrinsic device.

5.5.2 Ohmic Contacts

The contact resistance associated with an ohmic contact in a transistor device plays an important role in determining its performance, particularly at high frequencies. The source resistance, for instance, contributes directly to a FET's noise figure, and hence needs to be minimised. Similarly, for a high frequency Schottky diode, the parasitic ohmic resistance associated with the cathode contact impacts directly on the device's high frequency performance by reducing its cutoff frequency.

As outlined in Chapter 3, when the semiconductor in a Schottky contact is very heavily doped, the electrical characteristics of the junction change. Electrons can pass across the junction by means of *tunneling* rather than thermionic effects. As a consequence, the junction effectively becomes a low loss contact with insignificant diode-like characteristics. The resulting junction is known as an ohmic contact.

The conventional approach to contacting has been to use alloyed ohmic contacts. This makes it possible to form a very highly doped region near the surface. However, due to the non-uniformity of current flow through the contact, the small theoretical value of contact resistance cannot be realised. More advanced techniques, including ion implantation and MBE, are now more commonly applied.

Traditionally, an evaporated Au/Ge eutectic with a Ni over-layer was alloyed above the eutectic melting temperature. During this process, Ga vacancies formed. These were then filled by Ge donors. The Ni (or Pt) improved the melting of the eutectic film and increased the solubility limit in GaAs. A high contact resistance was often a limiting factor in terms of the resulting performance in high-speed semiconductor devices.

Ion implantation or diffusion can be used to create a highly doped region near the semiconductor surface. The doping concentration is limited by the impurity solubility limit and may reach $5 \times 10^{19} cm^{-3}$ for n-type, and $10^{20} cm^{-3}$ for p-type GaAs. Using ion implantation, the solubility limit may be reached near the surface.

MBE can also be used to achieve doping levels that are suitable for ohmic contact formation. Highly doped epitaxial layers can be grown between the active layer and the metal. This promising modification to the standard process is known as a heterojunction contact. The idea is to reduce the barrier height and contact resistance by introducing an intermediate semiconductor layer. It has been found empirically that the barrier height is close to two thirds of the energy gap. Consider, for example, a heterojunction of n^+ Ge on n^+ GaAs. The barrier height of Ge is about 0.5 eV versus that for GaAs of about 0.8 - 0.9 eV. Also, a very high doping concentration ($10^{20} cm^{-3}$) can be achieved in the Ge layer.

In summary, the key to achieving a low contact resistance is to have a very high carrier concentration in the contact region of the semiconductor material or, better still, to locate a narrow bandgap semiconductor layer between the metal and the GaAs in order that the barrier height be reduced. The older and less sophisticated technique of alloying metals into the GaAs is becoming less prevalent. Ohmic contacts to GaAs have yielded contact resistances of the order of $10^{-7}\,\Omega\, cm^2$.

6. CONSIDERATIONS FOR THE REALISATION OF EFFECTIVE MONOLITHIC MM-WAVE CIRCUIT LAYOUTS

6.1 Schematic optimisation consistent with good mm-wave layout practice

Mm-wave layout is radically different to layout at lower frequencies. At RF (low GHz frequency range), for example, the circuit elements are often crammed together to minimise die size. To an extent, coupling effects are considered, but only to a relatively minor extent. Circuit functions that need to be well decoupled in RFICs are physically separated on the die by careful selection of a suitable pad-out configuration. In the case of mm-waves, subtle coupling effects are much more of a concern throughout the circuit. Electromagnetic field coupling between adjacent elements is an issue due to the naturally occurring fields that radiate from transmission lines, for example. As a consequence, much iteration is necessary between design and layout. As the evolving schematic design demands a particular set of dimensions, it is typically found when the corresponding circuit is laid out that some circuit elements are located physically close together and suitable bends or dimension limits must be incorporated. These are then fed back into the schematic design simulations and the response re-tuned accordingly.

6.2 Consideration of foundry element limitations

As an example of the foundry element limitations, consider a spiral inductor. Spiral inductors are laid out as metal spirals (with suitable cross-overs/cross-unders) which are fabricated on top of the GaAs surface. In fact, due to its high resistivity, such spirals on GaAs are much more useful at high frequencies than similar structures fabricated on silicon due to the quality factors that can be achieved – it should be kept in mind that the substrate is the issue, not the spiral itself. However, the spirals are not ideal inductors, as they contain parasitic elements including capacitive coupling between adjacent turns and resistance. As a consequence, the equivalent circuit model for the spiral includes not alone the required series inductance but a number of these non-ideal parasitics. The net effect of the parasitics is that at high frequencies, the characteristics of the spiral change dramatically. At what is termed the self-resonant frequency, the effective reactance of the structure in fact changes from inductive to capacitive. As a consequence, in order that these parasitic effects, which are difficult to predict, do not significantly impact on the circuit response, the inclusion of spirals in mm-wave circuit designs demands that good design practice is observed. Typically, the self-resonant frequencies for the spirals supported by GMMT and modelled for the H40 process[3] are well below the mm-wave frequency regime. Thus, any mm-wave signals should be well decoupled at any part of the circuit in which such spirals are included. This will ensure that the unpredictable or unknown characteristics of the spiral at the mm-wave frequency do not become a factor in the circuit performance.

6.3 Probing considerations

This involves both RF and DC probing. In developing MMIC designs for volume applications, a key requirement is that the chips be capable of being wafer-probed. This then facilitates an effective screening at the wafer level, and the net result is that only parts that have passed appropriate tests are made available to customers. The alternative, which is often a reality for RFIC development due to the relatively cheap packaging and test costs, is that the die are separated, packaged and then tested prior to shipment to the customer. In the case of higher frequency MMIC developments, this approach is not wise for a number of reasons: the cost of packaging the die is expensive at mm-wave frequencies; the customer may require bare die for his application; the time lag between wafer fabrication and circuit test means that no meaningful feedback can be given to the wafer fabrication process engineers concerning issues with, or drifts in, circuit performance which may be due to process variation.

In the course of the layout development in this work, the DC inputs consisted of square bond pads that can also accommodate needle type bias injection probes for on-wafer probing. The RF input and output connections were laid out in a ground-signal-ground (GSG) format. The GSG probing format is the standard format used for high frequency wafer probing. The grounds on either side of the signal line provide reliable RF-ground connections to the circuit under test on the wafer; the connection from the metal chuck under the wafer to the wafer's backside metal does not provide a quality RF-ground. The packaging of such a connection involves bonding to the signal pad only. The two ground pads are usually left unconnected because in the package environment, a good RF-ground connection can be realised between the backside metal of the die and the package metal underneath.

The probes used for the purposes of evaluating the mm-wave MMIC designs developed during the course of this work are manufactured by GGB Industries[19], and are known as Picoprobes. A photograph of a typical RF probe with V-band (50 – 75 GHz) capability is shown in Fig. 2-4.

Figure 2-4. Photograph of a typical mm-wave Picoprobe for on-wafer measurement (source GGB Industries, Inc.[19]).

The coaxial connector on top of the probe is a V-type female connector and is known to work well up to beyond 67 GHz before moding becomes an issue. The long thin protrusion from the left side of the probe body contains the RF probe tips in a ground-signal-ground configuration. This probe arrangement is shown in the sketch in Fig. 2-5. The mechanical outline of a high frequency Picoprobe is shown in more detail in Fig. 2-6. The probe is mounted onto the probe station and must be carefully adjusted for planarity. The probe tips are very delicate and must be handled with care at all times.

6.4 Dicing/sawing considerations

To facilitate convenient dicing of the wafer after fabrication such that the individual die can be poked and packaged, adequate space must be allowed between the die on the GaAs through which the dicing or sawing tool can pass without damaging the fabricated circuits. This is accomplished by complying with the layout guidelines provided by the foundry. Essentially, these rules require a minimum amount of clearance between dice. The resulting channels between the dice are usually referred to as "scribe lines". If inadequate clearance were provided, two issues could arise. Firstly, the cutting tool could damage the circuits. Secondly, if the cutting tool were to pass through metal on the GaAs, this metal could get caught up in the tool and could cause erratic and poor quality cutting subsequently.

Figure 2-5. Ground-signal-ground probe arrangement (source GGB Industries, Inc.[19]).

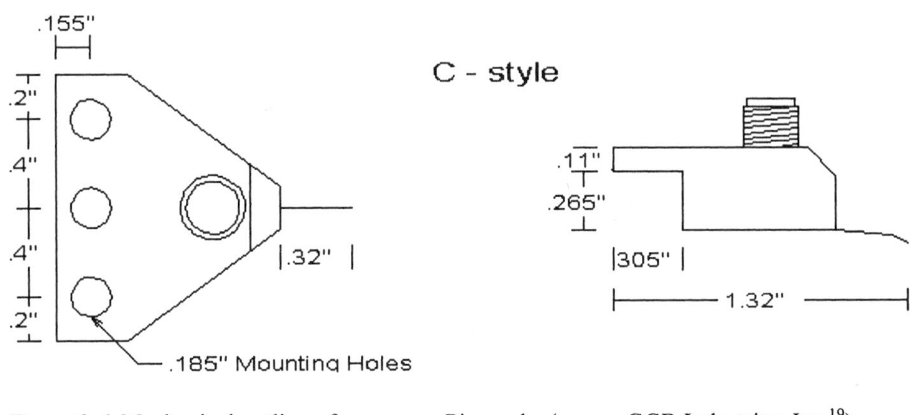

Figure 2-6. Mechanical outline of mm-wave Picoprobe (source GGB Industries, Inc.[19]).

6.5 Packaging impact on performance

At mm-wave frequencies, a key aspect of good design practice is that the I/O bond-wires are part of the design. In other words, the electrical effects of these wires are included in the circuit simulations and the performance is optimised with these effects in mind. This is particularly important at high frequencies where the inductive characteristics of the wires become very significant. In the case of the low noise amplifier developments described in Chapter 5, a particularly interesting broadband bond-wire compensation scheme, ideally suited to mm-wavelength applications, is introduced.

In some instances, especially at high frequencies, a pair of parallel wires is used instead of a single wire to reduce the effective inductance. However, this would generally require a larger bond pad to accommodate the two wires. Moreover, the characteristic of two adjacent wires is not as straightforward to predict as a single wire and in fact becomes a 3-D electromagnetic analysis problem which can be strongly dependent on the three dimensional geometry of the wires and will be somewhat different from assembly to assembly. As a consequence, in the course of this work, the more predictable single bond-wire approach was applied.

7. FUTURE TRENDS

An interesting development, which is currently provided by Kopin[20] as a commercial offering for HBT fabrication, is wafers with HBT transistors built in prior to shipment. The customer can then put these wafers through a wafer fabrication process where the interconnections between the transistors can be defined. This process is similar in nature to the gate arrays used in silicon technology. By forming the highly complex transistors prior to shipment to the circuit manufacturer, this approach can reduce the complexity of processing required at the fabrication facility responsible for 'wiring' the transistors together. This enables these customers to use mature, high yield 1 to 2 micron design rules to delineate and interconnect the transistors. This offers the potential for higher yields on high performance HBT circuits even while using lower cost, non-sub micron fabrication facilities. These higher yields and lower capital costs translate to lower circuit production costs.

Thickness and doping levels are the critical factors in the performance of high frequency devices. In addition for hetero-junction based devices, the interfaces must be abrupt or graded as appropriate to give the correct energy band interfaces. Therefore, both uniformity and reproducibility are important. Typically, 1% uniformities are needed to give good yields at the device level. The demand for high frequency devices and circuits has been growing. This has tended to push down costs, and a few companies have started high volume manufacturing. This is particularly true at the RF and low microwave frequency end of the application spectrum (in particular in the explosive wireless handset market), but the emergence of mm-wave applications will inevitably lead to volume manufacturing for these as well.

High frequency devices demand high quality semiconductor growth with very good control for thin layers being a particular requirement. The identification of the optimum epitaxial growth technique is a key issue to be resolved. To this end, the battle between MBE and MOCVD remains open. MOCVD certainly offers advantages in terms of cost, whereas the MBE offers outstanding growth control, which is of particular importance for very high frequency devices.

Chapter 3

HIGH FREQUENCY DEVICES

1. INTRODUCTION

In Chapter 1 of this work, a number of high frequency applications in the mm-wave range were described, and today these are driving developments of mm-wave integrated circuits for high volume manufacturing. In order that mm-wave integrated circuits be realisable, an obvious requirement is a circuit device, or range of devices, with the potential to function well at high frequencies. To facilitate the realisation of such a device or devices, a semiconductor material suitable for high frequency applications is necessary. Some pertinent semiconductor material options have previously been outlined in Chapter 2 of this work. These materials are typically III-V compound semiconductors. In this chapter, the physical structure and electrical characteristics of a number of high frequency devices are presented. The key differences in the structure and characteristics of these devices are described and explained.

These high frequency devices include a range of GaAs unipolar devices, extending from the very simple Schottky diode to much more sophisticated devices, such as the metamorphic HEMT, emerging from development laboratories in recent times. The GaAs hetero-junction bipolar transistor (HBT), which has shown great promise as a key high frequency transistor, is also discussed. The suitability of the various devices for high frequency applications will be analysed. Some brief thoughts on the potential for Si based devices for high frequency applications are also included at the end of the chapter.

2. HIGH FREQUENCY DEVICES

2.1 Background

Due to the relatively poor hole mobility of most semiconductor materials, high speed devices are typically unipolar (electron carrier based) in nature[1]. In recent years, this status has changed with the emergence of GaAs HBT devices and these will also be considered in this chapter. However, the vast majority of GaAs based devices are based on Schottky contacts, either as two-terminal diode type structures, or as three-terminal transistors with Schottky gates. These three terminal devices are field effect transistors (FETs) which are voltage controlled, as distinct from current-controlled bipolar type transistors used in Si technology. The most basic of the commonly used high frequency FETs is the metal-semiconductor FET or MESFET, generally fabricated on a GaAs substrate.

As will be outlined later in this chapter, hetero-junction FETs or HEMT devices can yield superior performance to conventional GaAs MESFETs. The HEMT exploits the wide band-gap properties of AlGaAs (doped) to generate a high density electron 'gas' in an adjacent undoped GaAs channel. Being undoped, the electron mobility in the channel is significantly greater than in the doped channel of the standard MESFET. This results in the HEMT having superior high frequency capability to the MESFET and a lower noise figure[14].

As has been mentioned in Chapter 2, GaInAs material has superior mobility to GaAs, and some hetero-junction type FETs incorporate this material to achieve performance superior to the conventional HEMT. GaInAs is not lattice matched to GaAs and thus the most common option is to accept a certain level of strain due to mismatch to a GaAs material system. This results in the pseudomorphic or p-HEMT. Another option is to use InP as the substrate material as this can be readily lattice matched to an appropriate mole fraction of GaInAs – see Fig. 2-3 in Chapter 2. This device is known as a lattice-matched HEMT. Being lattice matched to the substrate can make material growth more convenient but InP material is more expensive than GaAs, is more brittle and difficult to handle. Moreover, InP processing is less mature than that of GaAs, and the available wafers are smaller in diameter. As a consequence, significant effort has been expended in developing a GaAs substrate based equivalent to the lattice-matched

[1] Of course, the hole mobility in silicon is also relatively poor, and as a consequence, high frequency silicon technology tends to be based on n-MOS type devices which are also not constrained by minority carrier transport characteristics.

device. The result is the metamorphic HEMT, which has the channel arrangement of the InP substrate device and a complex layering arrangement to convert the GaAs substrate into a virtual InP one[21].

A GaAs based bipolar transistor, the HBT, has emerged in recent years[22]. This exhibits many of the traits of the conventional Si bipolar transistor, such as high transconductance and good current carrying capacity[23]. However, by an appropriate use of a hetero-junction between two semiconductor materials, the HBT can be manufactured with a heavily doped base and lightly doped emitter. This doping structure yields a device with useful current gain and low parasitic base resistance and emitter-base capacitance, key requirements for useful high frequency performance. The bipolar type device has a lower intrinsic *1/f* noise than a FET device, and as a consequence it is more suited to oscillator circuit designs than the FET. Due to its high current carrying capability and its high voltage gain, the HBT is also used extensively for power amplifier developments. As the commercial MMIC fabrication process available for use during the course of this work had a pHEMT device as standard, the HBT will be discussed only briefly in this chapter.

2.2 Schottky Diode

2.2.1 Introduction

The simplest GaAs based device is the Schottky diode. It consists of a metal - semiconductor junction, commonly referred to as a Schottky junction. The Schottky junction is an essential element of the MESFET family of transistors to be discussed later, but the diode itself plays an important role in a number of applications. In the analogue regime, the diode can be used in high frequency mixers, frequency multipliers and detectors (sensors). These circuits exploit the highly non-linear characteristics of the diode. In the digital world, the diode can be used for both level shifting and logic applications. A typical Schottky diode structure is shown in Fig. 3-1. Due to the critical role it plays in active monolithic GaAs devices, the Schottky diode will be discussed in some detail in this section.

2.2.2 Basic Theory

Any semiconductor can be characterised by a Fermi energy level, which generally lies between the valence and conduction bands -see Fig. 3-2. The location of the Fermi level relative to the valence and conduction bands is an indication of whether or not a semiconductor material is doped, whether or not n-type or p-type doping is involved, and to what extent the material is doped. If the material is n-type, then the Fermi level is closer to the conduction band and it is relatively easy for the electrons to acquire the additional thermal energy required to make the transition to the conduction band where they essentially become free to move around the lattice and conduct current.

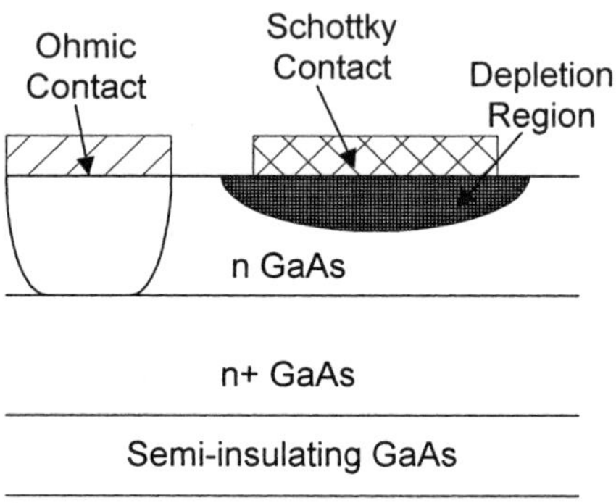

Figure 3-1. Schottky diode structure.

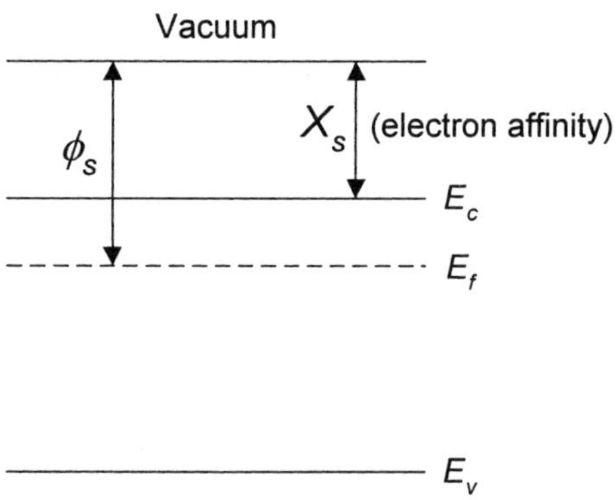

Figure 3-2. Semiconductor energy bands

The Fermi-Dirac distribution, which describes the probability function of finding an electron with an energy E (provided E is an allowed energy state), is given by:

$$F(E) = \left(1 + \exp\left[\frac{E - E_f}{kT}\right]\right)^{-1}$$

where k is Boltzmann's constant (1.38×10^{-23} JK^{-1}), T is the absolute temperature (K), and E_f is the Fermi energy level. The probability of finding an electron with an energy up to or equal to the Fermi energy level is 0.5 or 50%.

A straightforward analysis[4, 5], yields the following expressions, from which E_f can be deduced for an n-type semiconductor (a similar analysis can be applied for a p-type material)

$$E_i = \frac{1}{2}\left(E_c + E_v\right) + \frac{kT}{2}\ln\left(\frac{N_V}{N_C}\right)$$

$$n_i^2 = N_C N_V e^{\left(-E_g/kT\right)}$$

$$E_f \approx E_i + kT\ln\left(\frac{N_d}{n_i}\right)$$

where E_i is the Fermi level for an intrinsic semiconductor (i.e. undoped) and is normally located near the middle of the band-gap, E_c is the conduction band energy, E_v is the valence band energy, N_V is the effective density of states in the valence band, N_C is the effective density of states in the conduction band, E_g is the band-gap (equal to E_c - E_v), n_i is the intrinsic electron carrier concentration, and N_d is the doping density.

The electron affinity, X_S, represents the ease with which electrons can escape entirely from the semiconductor. Thermal effects and the associated statistical distribution of electron energies imply that a certain number of electrons can escape (on average). As the temperature is increased, this number will also increase as the electrons acquire more thermal energy.

Similarly, a work-function parameter, ϕ_m, can be defined for a metal. The work-function defines the energy differential between the vacuum level and the energy level associated with the metal electron cloud. The equivalent work function for the semiconductor is ϕ_s.

A Schottky barrier (electrostatic potential barrier) exists at the interface between a semiconductor and a metal. By virtue of thermal motion, some electrons will have sufficient energy to escape the semiconductor crystal.

As electrons leave the semiconductor, it becomes positively charged near its surface as ionised donors are left behind. The metal acquires a negative charge, which resides essentially on the surface. As this charge transfer occurs, band bending occurs which manifests itself as an electric field, which resists further electron transfer. A state of equilibrium is achieved when the tendency for electrons to move from the semiconductor to the metal is balanced by the reverse tendency due to the electric field.

The energy band diagram for the Schottky contact is shown in Fig. 3-3.

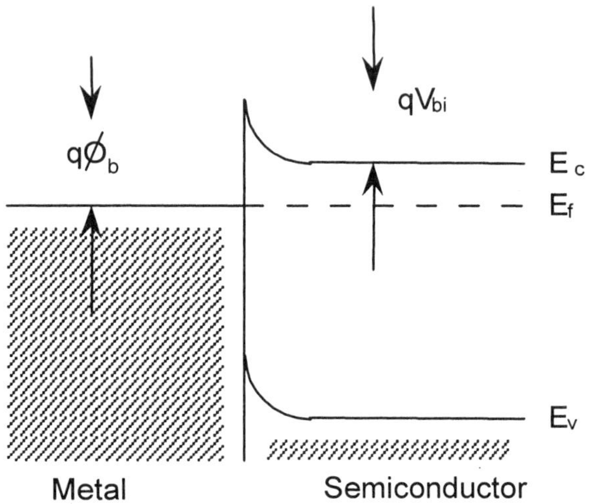

Figure 3-3. Energy band diagram for n-type semiconductor and a metal (no bias).

The potential difference V_{bi} (known as the *contact potential*), and the *barrier height*, ϕ_b, are defined by:

$$V_{bi} = \phi_m - \phi_s$$

$$\phi_b = \phi_m - X_s$$

From the energy band diagram in Fig. 3-3, the contact potential can also be expressed as:

$$V_{bi} = \phi_m - X_s - E_c + E_f$$

These terms are, at least in theory, functions of the semiconductor and metallic materials used. However, as will be outlined, this is not always the case in reality.

Due to the finite density of electrons in the semiconductor, a depletion region or space charge region (SCR) extends below the semiconductor surface. This depletion region exists where the bands are bending. The higher the doping concentration, the less the depletion region extends into the semiconductor, as the charge required to balance the charge at the metal surface can be provided by the ionisation of dopants to a lesser depth.

Applying the depletion approximation, a straightforward solution of the Poisson equation,

$$-\frac{\partial^2 V}{\partial x^2} \equiv \frac{\partial E}{\partial x} = \frac{\rho(x)}{\varepsilon_s}$$

yields the following expressions for the charge density, the electric field, and the potential respectively:

$$\rho = \rho(x) = qN_d$$

$$E(x) = -\frac{qN_d(A_0 - x)}{\varepsilon_s}$$

$$V(x) = -\frac{qN_d(A_0 - x)^2}{2\varepsilon_s}$$

where N_d is the doping concentration, A_0 is the SCR width with no bias applied across the junction, x is the distance from the surface, and ε_s is the semiconductor permittivity. These relationships are valid for $0 < x < A_0$.

The SCR width for an unbiased junction, A_0, can be deduced by setting $V(x)_{x=0} = V_{bi}$, and is given by:

$$A_0 = \sqrt{\frac{2\varepsilon_s V_{bi}}{qN_d}}$$

Thus, we see that the SCR width can be inferred directly from knowledge of the barrier height and the bulk semiconductor material parameters. When biased with a voltage V on the anode (with respect to the cathode), the above expression becomes

$$A_0 = \sqrt{\frac{2\varepsilon_s (V_{bi} - V)}{qN_d}}$$

The charge stored in the SCR is given by

$$Q = qN_d S A_o$$

where S represents the surface area of the diode. The corresponding depletion capacitance associated with this charge is given by

$$C_{dep} = \frac{dQ}{dV} = -S\sqrt{\frac{qN_d \varepsilon_s}{2(V_{bi} - V)}} \qquad (1)$$

Neglecting the negative sign, this depletion capacitance decreases as the junction is reverse biased, consistent with a parallel plate capacitor where the plates are moved further apart.

As mentioned above, the contact potential and barrier height do not strictly obey the theory expressing their values directly in terms of the metal work function, the semiconductor electron affinity and the gap between the conduction band and the Fermi level in the bulk semiconductor. The departure is due to surface states[16].

Surface states occur at the interface between the semiconductor and the thin oxide inevitably present at its surface. The oxide is so thin that electrons can easily tunnel through. The surface states, which influence the barrier height, are essentially dangling bonds at the semiconductor surface (i.e. a discontinuity in the ideally infinite semiconductor lattice) and the associated energy levels are distributed throughout the energy gap for covalently bonded materials like Si and GaAs.

Note that for ionic bonded materials like ZnS, there are no surface states within the energy gap and the standard theory applies.

For covalently bonded materials, the overall effect of the surface states is to reduce the impact of the metal selection on the resulting barrier height. The surface states are characterised by a 'neutral' level ϕ_o such that states below ϕ_o are neutral when filled with electrons, and states above ϕ_o are neutral when empty.

Measurements indicate that barrier heights depend on the method of surface preparation and often change with time. There is a definite correlation with the metal characteristics, but this correlation is weaker than would be suggested by the basic theory.

The Schottky effect should also be mentioned at this point. This refers to the image-force-induced lowering of the potential energy for charge carrier emission when an electric field is applied to the Schottky junction[4]. Essentially, when an electron is approaching the metal (from the semiconductor), it induces positive charge on the metal surface. This results in an attractive force between the electron and the metal, which reduces the work that must be done to bring the electron to the metal surface. The net result is a slight rounding of the semiconductor conduction band near the metal surface and a small reduction in the barrier height.

2.2.3 Schottky Diode I/V Characteristics

The current flow in a metal - semiconductor Schottky structure is limited by:

(a) Drift and diffusion through the SCR for low doping density material

(b) Thermionic emission for highly doped material

The energy band diagram is modified by the application of bias, as indicated in Fig. 3-4 and Fig. 3-5.

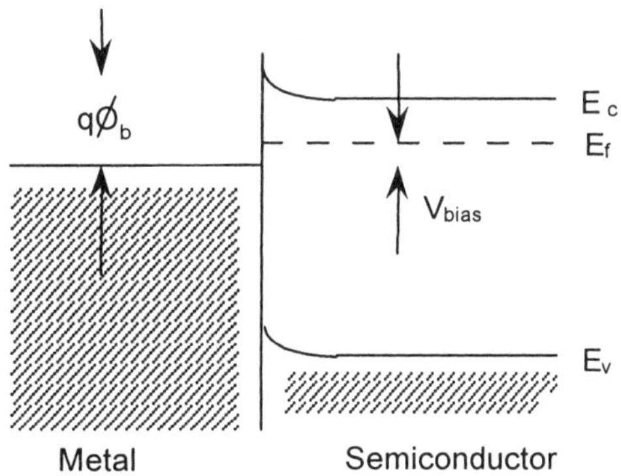

Figure 3-4. Energy band diagram for Schottky contact under forward bias.

The biased Schottky junction analysis is given excellent treatment elsewhere, e.g. Sze[4], and only a few pertinent points and the key results will be mentioned here.

The thermionic emission theory was developed initially by Bethe[24] and is based on the assumption that the barrier height is much greater than kT. Essentially, the current flow from the semiconductor to the metal is determined from a summation of the electrons which have sufficient energy to get over the barrier. This in turn is a function of the electron-energy distribution (Fermi-Dirac) function presented earlier, and a density of states function. This current flow is found to increase exponentially with an applied forward bias (metal biased positively with respect to the semiconductor). With no applied bias, there is no net current flow across the junction, and the current from the semiconductor to the metal is cancelled by an equal current flowing from the metal back to the semiconductor.

However, when a forward bias is applied, the tendency for current flow from the semiconductor to the metal increases. At the same time, the barrier preventing current flow from the metal to the semiconductor does not change, and hence the associated current is constant. Thus, it is clear that, with the application of bias, a net non-zero current flow exists. This net current is almost exponentially related to the applied bias.

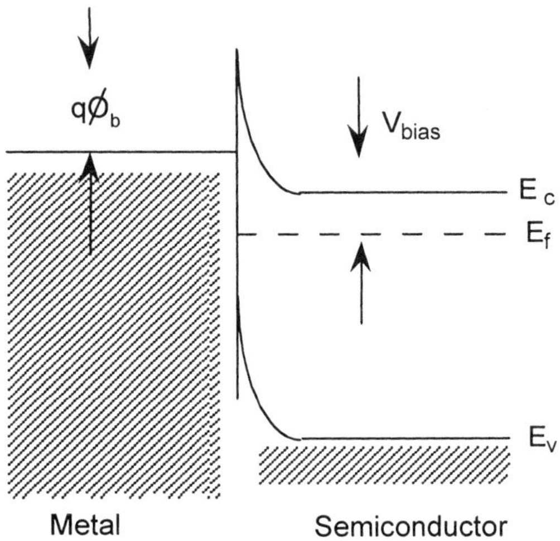

Figure 3-5. Energy band diagram for Schottky contact under reverse bias.

An alternative analysis, based on diffusion theory, was developed by Schottky[25]. This approach applies the conventional drift/diffusion theory and arrives at a solution which is very similar in form to the thermionic result of Bethe. This result is appropriate for non-highly doped semiconductor junctions.

The two results have the same exponential bias dependence. However, the saturation currents derived from the two analyses are not the same and exhibit different dependencies on the voltage and temperature. A synthesis of the thermionic and diffusion approaches was developed by Crowell and Sze[26]. This theory essentially arrives at a solution, again similar in form to the Bethe and Schottky results, but which can be simplified to either of the above results, depending on the level of doping.

The key result of Crowell and Sze can be summarised as follows. The model predicts the following for the junction current:

$$j = j_o \exp\left(\frac{qV}{\eta kT}\right)\left[1 - \exp\left(-\frac{qV}{kT}\right)\right]$$

$$j_o = A^{**}T^2\exp\left(-\frac{q\phi_b}{kT}\right)$$

where A^{**} is the modified Richardson constant and V is the applied bias voltage. For GaAs, $A^{**} = 4.4\text{x}10^4\ Am^{-2}\ K^{-2}$.

The ideality factor, η, is related to the voltage dependence of the barrier height. Typical values for η lie in the range 1.02 to 1.6. This ideality factor is a measure of how ideal the diode characteristic is, and the greater its value (above its ideal value of unity), the less non-linear the diode is. In terms of the current voltage characteristic of a Schottky diode, a large ideality factor is associated with a 'soft' turn on of the current versus voltage.

The saturation current j_o for a Schottky junction is much greater than the corresponding current for a p-n junction (as the Schottky's barrier height is lower), and the cutin voltage is smaller. This result has the same form as that of Bethe's thermionic model.

The thermionic model is valid when the mean free path (average distance travelled by an electron between collisions), λ, is much greater than the distance over which the barrier height decreases by kT/q (the barrier height varies more rapidly in a highly doped semiconductor as the associated ionised charge density is higher). The mean free path is given by

$$\lambda \approx \frac{\mu}{q}\sqrt{1.5kTm}\ ,$$

where m is the carrier effective mass.

For GaAs at room temperature, this indicates that the model is valid for $N_d > 10^{14}\ cm^{-3}$, assuming $V_{bi} = 0.7V$, $\mu = 0.5m^2V^{-1}s^{-1}$ as shown by Shur[18]. The levels of doping typically used in GaAs semiconductor devices fall within this range, and as a consequence the thermionic model is normally used (but of course, the more complex Crowell and Sze model is also valid).

2.2.4 Tunnelling

In a highly doped semiconductor, the depletion region becomes so thin that electrons near the top of the barrier can tunnel through. This is referred to as *thermionic field emission*. The I/V characteristic is determined by competition between the thermal activation and tunneling and is given by:

$$J = J_{stf}\exp\left(\frac{qV}{E_o}\right)$$

In GaAs, thermionic field emission occurs roughly when $N_d > 10^{17}\ cm^{-3}$ at 300K, or $10^{16}\ cm^{-3}$ at 77K. For Si, these values are about a factor of four times larger.

In degenerate semiconductors (where the Fermi level in the bulk exceeds E_c), especially with small electron effective mass, electrons can tunnel through the barrier near the Fermi level. This is referred to as *field emission*. The IV characteristic has a similar form:

$$J = J_{sf} \exp\left(\frac{qV}{E_{00}}\right)$$

It can be shown[4] that the tunnelling current increases exponentially with $\sqrt{N_d}$. Tunnelling is also greater at reduced temperatures.

The effective resistance of the Schottky barrier in the field emission regime is low. As a consequence, the metal – n^+ barrier is used to make efficient ohmic contacts. Such an ohmic contact plays a key role in the Schottky diode in facilitating an efficient contact to the cathode. It also provides the source and drain contacts for the field effect transistors to be discussed shortly.

In conclusion, the IV characteristics of a Schottky diode are similar to those of a p-n junction, being exponential in form. The saturation current is higher for a Schottky diode. The current flow mechanisms vary significantly depending on the doping concentration and material mobility. At one extreme, the current is limited by diffusion and drift processes; at the other extreme, the carriers can tunnel through the barrier near the Fermi level. Despite these mechanism differences, the resulting expressions relating the current and applied voltage are similar, with the primary differences involving the temperature dependencies of the key parameters.

2.2.5 Schottky Diode Model

For the purposes of circuit design using a Schottky diode, a suitable model is required. The type of model most suitable for the simulation is dependent on the application. In some CAE design tool packages, such as Agilent-Eesof's ADS and Libra[27], the Schottky diode is implemented as a special case of a p-n diode.

In terms of equivalent circuits, a commonly used Schottky diode model is shown in Fig. 3-6. The circuit contains primarily a voltage dependent resistance R_d, a voltage-controlled capacitance C_{dep}, and a series resistance R_s. The series resistance is associated with the resistance of the undepleted semiconductor material in addition to the contact resistances of the ohmic contact, and is generally small. The voltage-controlled capacitor is associated with the charge stored in the depletion region, and is given by Eq. (1). Under reverse bias, the depletion region expands and the charge also increases. The variation in stored charge and depletion region depth contributes directly to a voltage dependent capacitance. We have a voltage-controlled resistance, R_d, in parallel with the capacitor. This equivalent circuit component represents the resistance associated with the depleted SCR. As the diode is reverse biased and the SCR extends further away from the Schottky contact, the value of R_d increases. The series resistance is a non-ideality, which complicates the behaviour of the diode device. Both R_d and C_{dep} are *non-linear* elements. This means that their values are not constant, but vary with an instantaneous voltage value. Such non-linearities can make the analysis of diode based circuits difficult. A small signal frequency-domain analysis may not be adequate, and a time-domain analysis may be necessary. However, these same non-linearities can provide very useful benefits in terms of circuit functionality, as will be seen later.

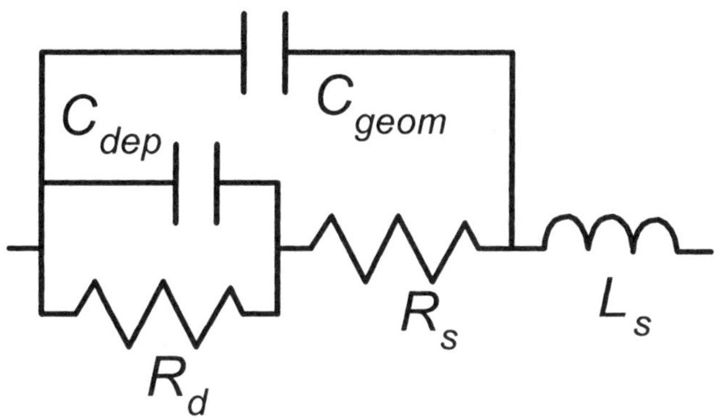

Figure 3-6. Schottky diode equivalent circuit model.

Depending on the doping profile in the Schottky diode material structure, the device's non-linear conductance or capacitance may dominate. A non-linear conductance is suited to mixing or detecting applications, whereas a non-linear capacitance is more suitable for harmonic generation (frequency multipliers). A capacitance due to the geometry of the device is represented by C_{geom}. This accounts for the capacitive coupling across the device due to its physical structure. The capacitance is a constant. A series element, L_s, accounts for the effective inductance associated with the device connections.

2.3 MESFET

2.3.1 Introduction

A field-effect transistor (FET) is a particular transistor type where the current flow is controlled by an applied voltage. A Metal Semiconductor Field Effect Transistor (MESFET) is a particularly important transistor for high frequency circuit design. Its cross-section is shown in Fig. 3-7. The structure of the MESFET is similar to that of a silicon based MOSFET, which is widely used in lower frequency electronic circuits. The key difference in the structures is that the MESFET gate lies directly on top of the semiconductor, whereas there is an intervening layer of dielectric in the MOSFET (SiO_2). In this context, the MESFET is quite similar to a JFET. As a consequence, the possibility of non-zero gate current is a characteristic of MESFET devices. Generally speaking, gate current is not desirable, and this places constraints on the biasing of the device in a given application. The MESFET is the most basic GaAs FET type transistor. A GaAs based MOSFET type transistor would be a very attractive device, but unfortunately, has proved very difficult to realise due to the absence of a high quality native oxide. Some progress has been made in the development of a GaAs MOSFET and devices with promising high frequency capability have been reported[28]. These use Ga_2O_3 as the gate dielectric material.

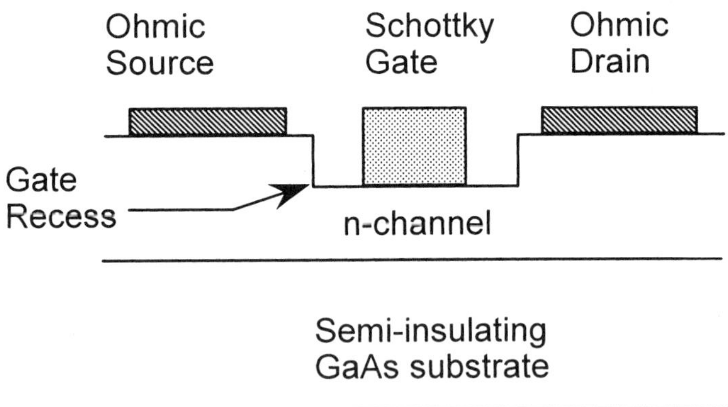

Figure 3-7. MESFET cross-section.

The relative characteristics of Si-based MOSFETs and GaAs-based MESFETs have been studied by Shur[17]. A comparison of the transconductance and drain current characteristics of typical devices is shown in Fig. 3-8.

The MESFET has been the transistor workhorse of high frequency MMIC circuits. It is the transistor device that has made monolithic microwave and mm-wave amplifiers and oscillators possible. The simple material structure of the MESFET has been refined over the years, resulting in more exotic devices like the HEMT, which are capable of providing superior high frequency performance. Due to its importance, the MESFET characteristics will be discussed in some detail in this section. Much of the detail provided on the MESFET will also apply to the MESFET variants to follow.

2.3.2 Basic MESFET Operation

The MESFET is typically operated with a positive drain-source voltage, V_{ds}, and the drain-source current, I_{ds}, is controlled by the gate-source voltage, V_{gs}. Both the gate-source and gate-drain junctions are Schottky diodes, with a space charge region (SCR) extending beneath the gate. This SCR constricts the channel region through which I_{ds} can flow. When the gate is reverse biased (with respect to both source and drain), the channel is further constricted as the SCR width increases, and I_{ds} is reduced. When the gate is slightly forward biased, the SCR width is reduced and I_{ds} will increase. As a consequence, the current flow is modulated by the voltage applied to the gate.

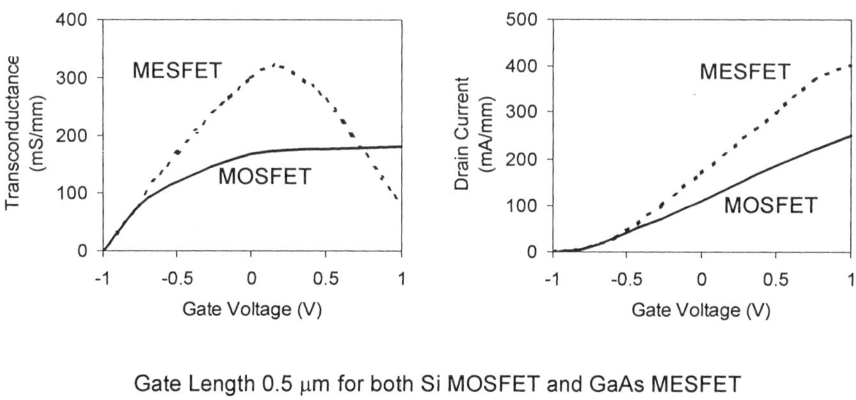

Figure 3-8. Comparison between MOSFET and MESFET characteristics.

Let us examine this operation in more detail: firstly, consider the situation where V_{ds} is relatively small, and the device is in its *linear* region. In this situation, for a given gate source voltage, and a given drain source voltage (assumed positive), the depletion region will be somewhat deeper at the drain end of the channel - see Fig. 3-9.

Now, consider the current flow in the case where the depletion region does not exist, and the channel available for current flow corresponds to the full channel depth, A. In this situation, the conductance of the channel is given by

$$g_0 = \frac{Conductivity \times Area}{Length} = \frac{qN_d \mu_n WA}{L_g}$$

However, due to the existence of the Schottky contact and its associated depletion region, the actual channel conductance is smaller. At a given position x in Fig. 3-9, the depletion depth, $d(x)$ is given by

$$d(x) = \sqrt{\frac{2\varepsilon_s(\phi_b - V_{gs} + V(x))}{qN_d}}$$

where $V(x)$ is the channel potential relative to the source side. This expression is identical in form to that derived earlier for the Schottky contact and is valid for small V_{ds} (gradual channel approximation).

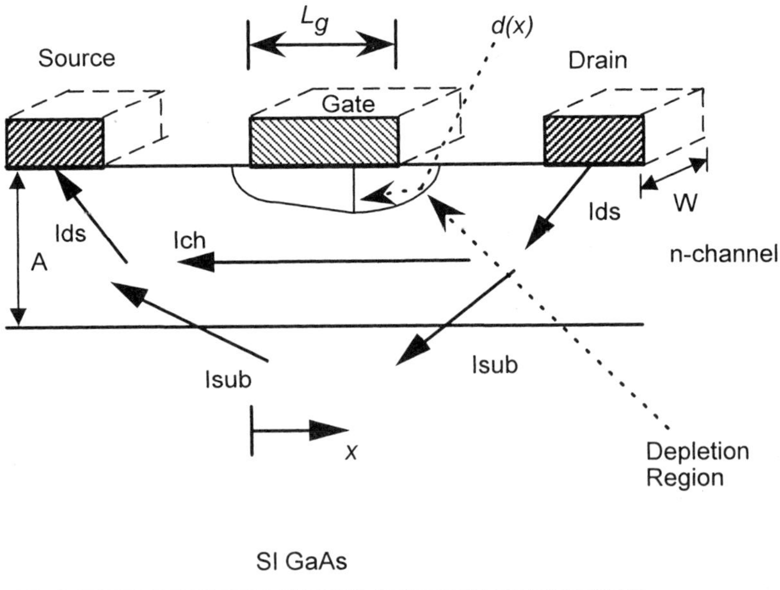

Figure 3-9. MESFET depletion region.

The channel conductance is reduced to zero when the reverse bias on the gate is sufficient to deplete the channel fully at the source end. This condition is satisfied when

$$d(0) = \sqrt{\frac{2\varepsilon_s(\phi_b - V_T)}{qN_d}} = A,$$

and the corresponding V_{gs} is known as the *threshold* voltage, V_T. The *pinchoff* voltage, V_p, is similarly given by

$$V_p = \phi_b - V_T$$

and is also clearly found to be

$$V_p = \frac{qN_d A^2}{2\varepsilon_s}$$

Now, utilising our knowledge of the variation of the actual channel conductance, g_{ds}, with location along the channel, we can derive an expression for the total channel current. The assumption made is that the channel current, I_{ds}, is constant along the channel length. The expression developed is valid for small V_{ds}.

Consider an infinitesimally small section of the channel, at position x, and length δx. Assuming a constant mobility along the channel, the voltage drop across this section is simply given by the product of the current and the resistance associated with the small section of the channel:

$$\delta V = I_{ds} R = I_{ds} \frac{\delta x}{qN_d \mu_n W[A - d(x)]}$$

By multiplying by the denominator of the right hand side, the above can now be rewritten as

$$qN_d \mu_n W[A - d(x)]\delta V = qN_d \mu_n WA\left[1 - \sqrt{\frac{\phi_b - V_{gs} + V(x)}{V_p}}\right]\delta V = I_{ds}\delta x$$

and can now be integrated on both sides with respect to x over the full channel length, yielding

$$qN_d W\mu_n A\int_0^{V_{ds}}\left[1-\sqrt{\frac{\phi_b - V_{gs} + V(x)}{V_p}}\right]dV = I_{ds}L_g$$

The above integration is straightforward and, importantly, requires no assumptions with regard to the shape of the *V(x)* curve (because *V(x)* is a variable of integration). On solving the integration, the following expression is easily found for the channel current:

$$I_{ds} = \frac{qN_d\mu_n WA}{L_g}\left(V_{ds} - \frac{2}{3}\left[\frac{(\phi_b - Vgs + V_{ds})^{3/2} - (\phi_b - V_{gs})^{3/2}}{V_p^{1/2}}\right]\right)$$

This reduces to the fundamental equation of the field effect transistor:

$$I_{ds} = g_0\left(V_{ds} - \frac{2}{3}\left[\frac{(\phi_b - Vgs + V_{ds})^{3/2} - (\phi_b - V_{gs})^{3/2}}{V_p^{1/2}}\right]\right) \qquad (2)$$

where the first term, $g_0 V_{ds}$ is the current in the full channel with no SCR, and the remaining terms represent the reduction in that current due to the existence of that SCR.

Applying Eq. (2) for a FET with the parameters $\phi_b = 0.6V$, $V_p = 5V$, the resulting current as a function of the applied biases, normalised to g_0 is represented in Fig. 3-10. A few points are worthy of note here. Firstly, for all V_{gs} traces, the current increases approximately linearly with V_{ds} for low values of V_{ds}. As V_{ds} increases further, the increased reverse bias of the gate-drain junction results in a channel that is increasingly choked at the drain end; hence the reduced slope in the I/V characteristic. Secondly, a lower current is associated with a more negative V_{gs}. This is consistent with the available channel depth as a whole, and particularly at the drain end, being reduced by the increased depletion width. Thirdly, for V_{gs} = -3V, the current trace initially increases with increasing V_{ds}. However, when V_{ds} exceeds 1.4V, the predicted current actually *decreases* for further V_{ds} increases. This is counter-intuitive, and is not borne out by measurement.

The problem here is that the above analysis, and Eq. (2) in particular, is only valid for small values of V_{ds}. It makes the gross assumption that the carrier mobility is constant, and that the carrier velocity can increase without limit. As has already been outlined, this is not the case. The carrier velocity has a maximum value, and in fact for GaAs and other compound semiconductors, further increases in the electric field can lead to a reduction in the carrier velocity.

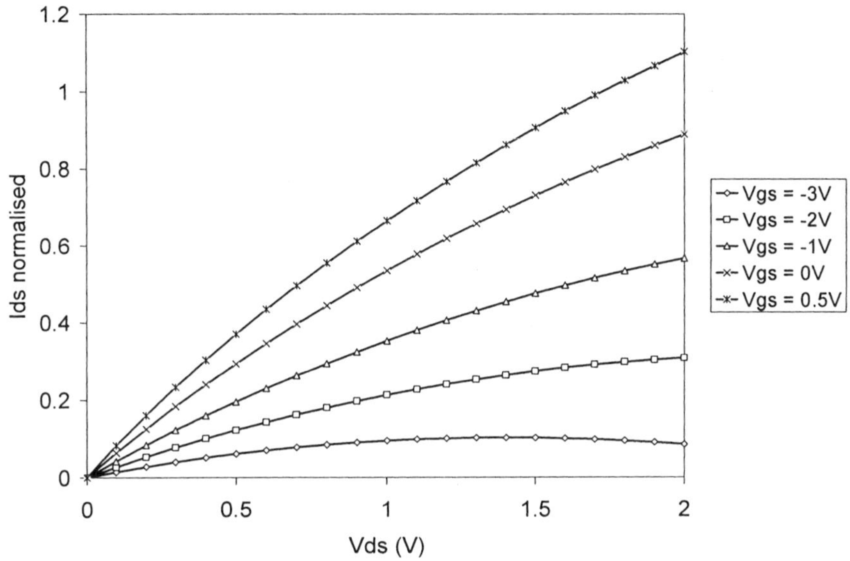

Figure 3-10. Normalised MESFET I/V characteristic based on Eq. (2).

A qualitative description of the I/V characteristics of a real device now follows. As V_{ds} increases, the depletion width at the drain end of the channel increases and the channel width there is reduced. Thus, in order for the fixed current to flow across the channel, the carrier velocity must become greater as the channel width decreases (this is consistent with the electric field increasing in this region as well). The velocity field characteristics of semiconductors have already been discussed. It is an unavoidable fact that the carrier velocity cannot increase indefinitely. For GaAs, the carrier velocity will saturate at a value of approximately $1x10^7$ cms^{-1}. Thus, as V_{ds} continues to increase, a situation will be reached where the carrier velocity saturates and cannot increase any further. Further increases in V_{ds} do not lead to increased carrier velocities. Instead, a complex situation arises where, in order to maintain a fixed current in a further reduced available channel, the effective charge density must increase. A charge domain (referred to as a Gunn domain) develops. In simple terms, negative charge accumulates as the velocity saturates where the SCR width continues to increase (i.e. more electrons can flow into than out of this region). Further on towards the drain end, where the channel begins to open again, a positive charge develops. This domain formation is a feature of all field effect transistors, whether they are fabricated on GaAs, Si, or some other semiconductor material. However, due to their unusual velocity – field characteristics, the charge domain formation is more complex in compound semiconductor based FETs.

The net outcome of this behaviour is that the I/V traces saturate with increasing V_{ds}. This is why the simple model, represented by Eq. (2) above, and represented graphically in Fig. 3-10, is valid only for small V_{ds}. As V_{ds} increases above a value, denoted $V_{ds(sat)}$, the velocity saturation effects come into play and the current characteristic deviates from the simple representation in Fig. 3-10. A value for $V_{ds(sat)}$ can be crudely estimated from the following expressions for the saturated electron velocity:

$$v_{sat} = \mu E_{sat} = \mu V_{ds(sat)} / L_G$$

such that

$$V_{ds(sat)} = v_{sat} L_G / \mu$$

As will be discussed, for high frequency performance to be possible, short transit times under the gate region are necessary, thereby demanding short gate lengths, L_G. For reasonable mm-wave performance, gate lengths of 0.25μm are typical. Applying the GaAs parameters from Table 2-1 in Chapter 2, the resulting value obtained for $V_{ds(sat)}$ is 0.05 – 0.1 V. This suggests that for a typical mm-wave GaAs FET, the expression in Eq. (2) remains valid only up to a V_{ds} of about 0.1V. Beyond $V_{ds(sat)}$, the I/V characteristic rounds off and the current goes into saturation. The domain, which develops underneath the drain end of the SCR, plays a key role in maintaining an approximately constant current for further increases in V_{ds}. In fact, as V_{ds} continues to increase, the domain charge increases and the point under the SCR where the electron velocity saturates moves slightly towards the source. This corresponds to a slightly more open channel, and hence is consistent with a small increase in the transistor's saturation current (see Fig. 3-11). This effect is known as channel length modulation.

There are two flavours of MESFET, enhancement mode (E type) and depletion mode (D type). The E-type MESFET requires a positive V_{gs} to enable drain current to flow, whereas in the D-type device, current flows for negative V_{gs} down to the threshold voltage V_T. These two device variants are outlined in Fig. 3-11. Generally speaking, for $V_{gs} = 0$, the depletion region in the E-MESFET extends all the way across the conducting channel, whereas in the D-MESFET, the channel remains partly open. An enhancement mode characteristic can be achieved in a device either by making the channel thinner than would be appropriate for a D-MESFET, or else by using a lower doping concentration in a channel of similar thickness. The significance of this is that a D-MESFET often requires both negative and positive supply voltages for useful circuit operation – this can be avoided through the use of self-biasing, but this is not always desirable for optimum circuit performance. A single supply is always sufficient for an E-MESFET.

It can also be noted from Fig. 3-11 that for a fixed V_{gs}, the saturation current continues to increase slightly with increased V_{ds}, as discussed above.

2.4 Modern FET Variants

In recent times, as compound semiconductor processing capability has advanced, more complex material structures have become possible. This has been driven in particular by MBE and MOCVD epitaxial growth techniques, which make possible the growth of precise thin high quality layers of material with excellent control – see Chapter 2. This capability has made viable a range of hetero-junction based FET devices which are similar to the conventional MESFET but offer superior high frequency capability.

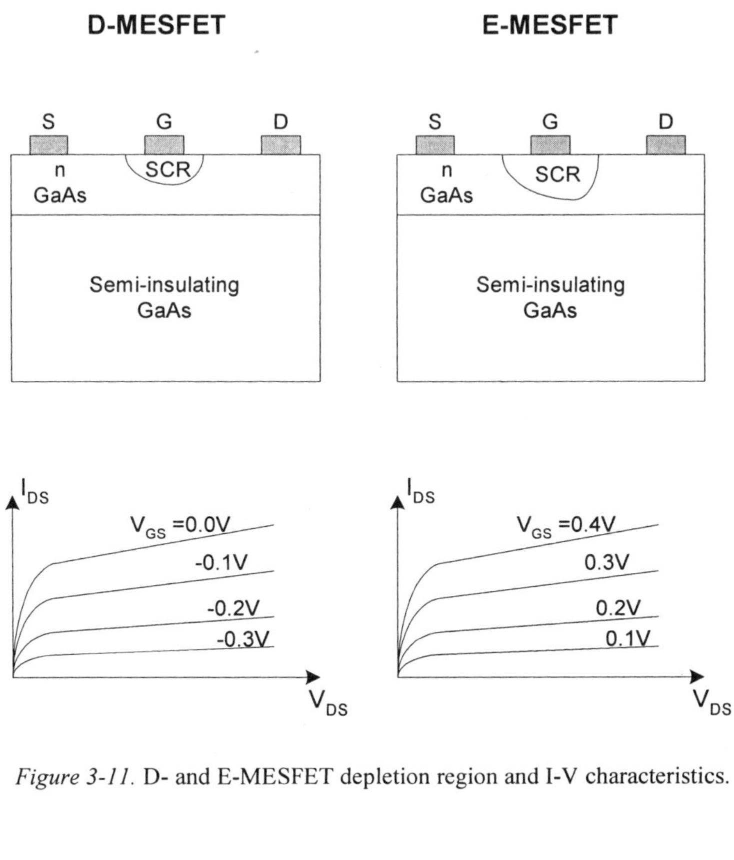

Figure 3-11. D- and E-MESFET depletion region and I-V characteristics.

2.4.1 HEMT

The HEMT (High Electron Mobility Transistor) is a more complex variant of the MESFET, which yields superior performance at high frequencies. In the MESFET, the GaAs channel is doped and the electron mobility is inevitably less than that of undoped GaAs. In the HEMT, a hetero-junction is used to inject carriers (from a doped wide band-gap AlGaAs) layer into a (narrow band-gap) undoped GaAs channel, thereby resulting in a higher mobility and hence better high frequency performance. The associated elimination of undesired scattering events also improves the noise performance as the scattering occurrences are essentially unpredictable and random and hence contribute to noise. Due to the nature of the band structure of the hetero-junction, the electrons in the undoped GaAs are confined to a very thin region. This region of GaAs with a very high electron density is known as a two-dimensional electron gas (2DEG). The band bending which occurs in a HEMT hetero-junction is illustrated in Fig. 3-12.

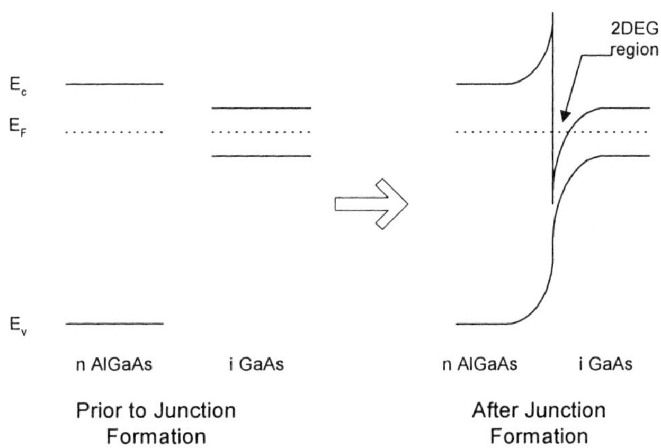

Figure 3-12. Formation of band structure in AlGaAs –GaAs hetero-junction HEMT.

The cross section of a typical HEMT device is shown in Fig. 3-13. In the example in Fig. 3-13, the doping in the wide band-gap AlGaAs is introduced in a very small thickness. This is made possible by the use of MBE as the epitaxial growth technique. This thin doping arrangement is known as delta doping. Alternatively, a thicker slice of n-AlGaAs material could be doped. The delta doping method gives better uniformity and control over the device characteristics.

The operation of the HEMT device is very similar to that of a MESFET. The difference here is that the channel is separated from the gate, and the channel depth is very small. In much the same way as in the MESFET, the voltage on the gate influences the conduction properties of the channel (2DEG).

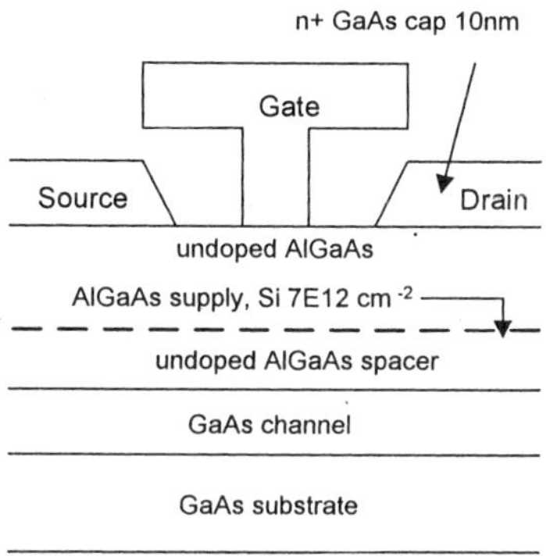

Figure 3-13. Cross-section of HEMT device.

A typical drain current versus gate voltage for a 200μm HEMT device ((0.15μm gate length), from Angelov et al[29], is shown in Fig. 3-14.

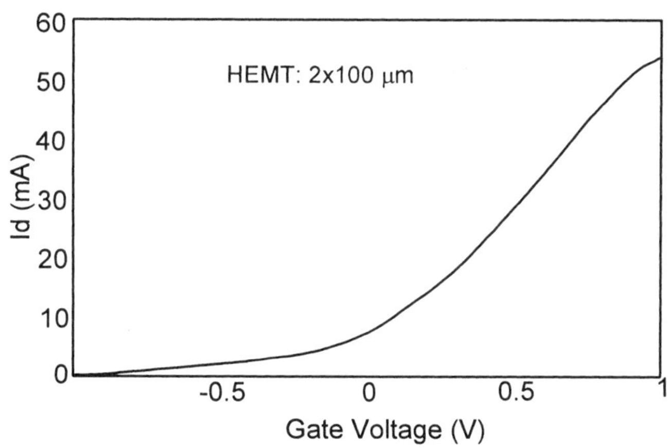

Figure 3-14. Typical HEMT (200 μm, 0.15 μm gate length, V_{DS} =2V) I - V characteristic (source Angelov et al[29], © 2004 IEEE).

The corresponding transconductance trace[29] is shown in Fig. 3-15.

Based on these Figures, the following figures of merit are estimated for this HEMT device:

I_{max} = 270 mA/mm; g_m = 300 mS/mm

These figures of merit are usually normalized per mm so that the performance of different devices can be compared fairly for a standard gate width of 1mm.

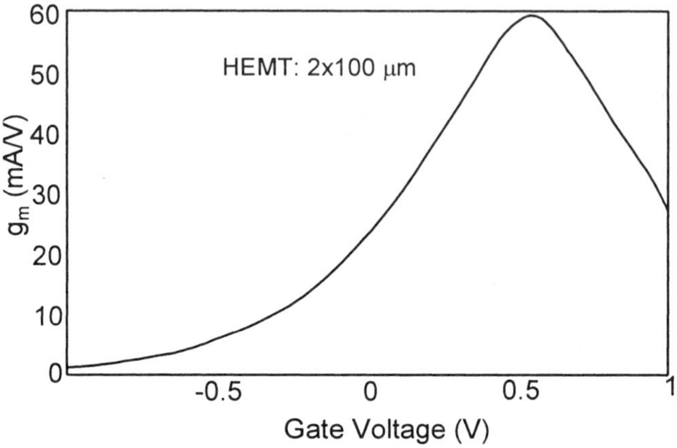

Figure 3-15. Typical HEMT (200 μm, 0.15 μm gate length, V_{DS} =2V) g_m - V characteristic (source Angelov et al[29], © 2004 IEEE).

2.4.2 Pseudomorphic HEMT

The pseudomorphic HEMT (pHEMT) is a more complex version of the conventional HEMT, which exploits the excellent carrier transport properties of InGaAs. The layering structure of a pHEMT is shown in Fig. 3-16. This device is similar in nature to the regular HEMT, in the sense that the hetero-junction structure forces electrons from the doped wide band-gap AlGaAs to be injected into the undoped narrow band-gap InGaAs.

Most commercial high frequency foundry processes today are based on the pHEMT device. Excellent mm-wave performance has been reported from many sources for circuits designed using transistors with gate lengths of 0.25 μm and less. For example, in Cameron et al[30], details on devices with a f_t of 121 GHz and a f_{max} of 157 GHz were published – see Section 2.6 of this chapter for definitions of these parameters.

The measured drain current versus drain voltage characteristic for a pHEMT (L_G = 0.2 μm, gate width = 160 μm) is shown in Fig. 3-17. This device has demonstrated excellent high frequency capability[31]. The 0.2 μm gate length of this device is different to that of the HEMT whose characteristics were presented in Figs. 3-14 and 3-15, and hence it is difficult to draw conclusions from a comparison.

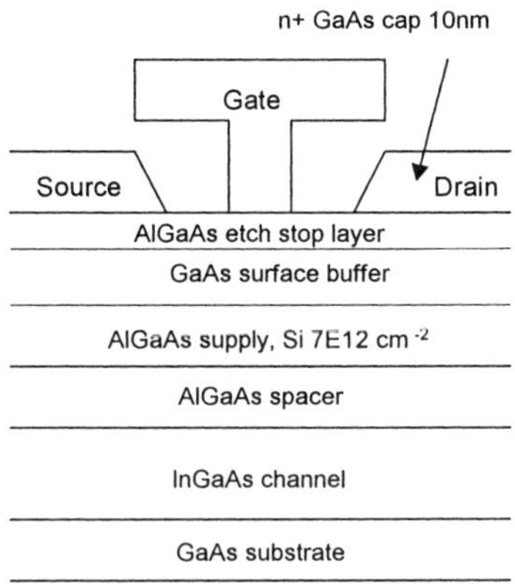

Figure 3-16. Cross-section of pHEMT device.

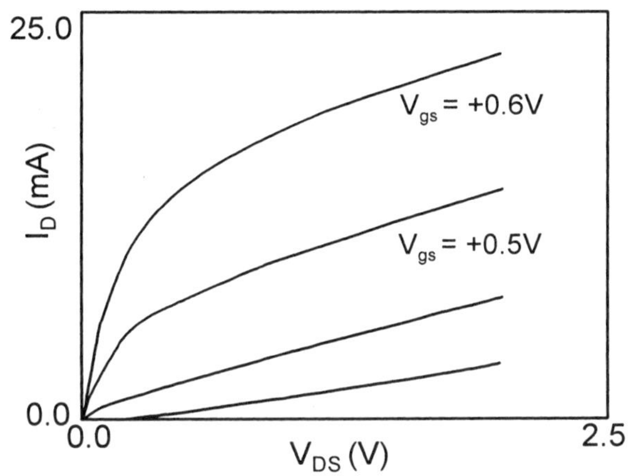

Figure 3-17. Measured pHEMT (160 μm, 0.2 μm gate length) I - V characteristic (source Ishikawa et al[31], © 2004 IEEE).

A typical drain current versus gate voltage for a 200 μm gate width pHEMT device (0.15 μm gate length), from Angelov et al[29], is shown in Fig. 3-18. This device has the same overall gate dimensions as the HEMT in Fig. 3-14. It is clear from a comparison with Fig. 3-14 that the pHEMT is inherently a higher current device than the conventional HEMT (400 mA/mm versus 270 mA/mm). The transconductance trace for the same pHEMT device is shown in Fig. 3-19. Again, it can be seen that the pHEMT transconductance (375 mS/mm) exceeds that of the HEMT in Fig. 3-15 (300 mS/mm).

2.4.3 Lattice-Matched HEMT on InP

The channel material in a pHEMT is not lattice matched to the underlying GaAs substrate. A more ideal material growth scenario is obtained when the epitaxial structure is grown on an InP substrate instead. Under these circumstances, it is found that a hetero-structure between InAlAs (doped wide band-gap material) and InGaAs yields good performance.

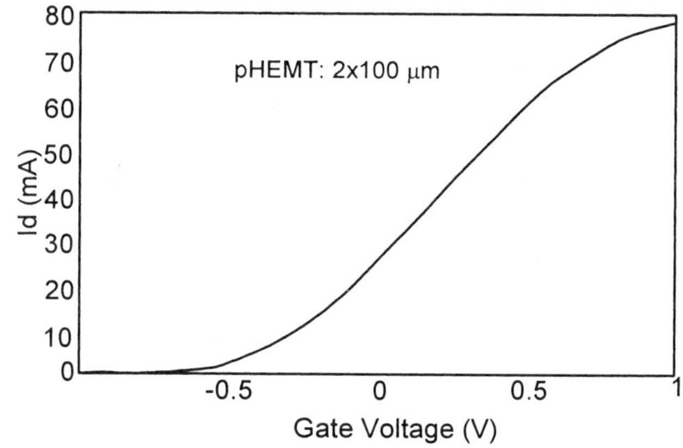

Figure 3-18. Typical pHEMT (200 μm, 0.15 μm gate length, V_{DS} =2V) I - V characteristic (source Angelov et al[29], © 2004 IEEE).

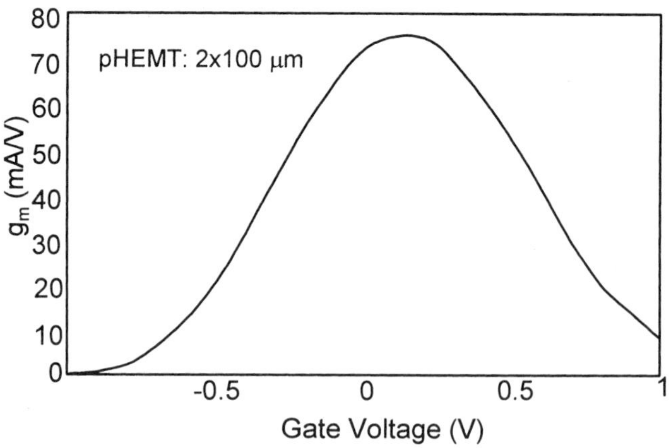

Figure 3-19. Typical pHEMT (200 μm, 0.15 μm gate length, V_{DS} =2V) g_m - V characteristic (source Angelov et al[29], © 2004 IEEE).

This material system has attracted much attention due to the large conduction band discontinuity between these materials (low leakage current and high breakdown voltage). The cross section of such a lattice matched HEMT (LM HEMT) is presented in Fig. 3-20. A good lattice match exists between InP and $In_{0.53}Ga_{0.47}As$ (see Fig 2-3 in Chapter 2).

Excellent results have been reported for circuits based on the LM HEMT. For example, very promising W-band amplifiers have been developed[32] using devices with a 0.25 μm gate length. A high sheet charge density of $3.5\times 10^{12}\ cm^{-2}$ and an electron mobility of greater than 10,000 $cm^2V^{-1}s^{-1}$ have been reported[33].

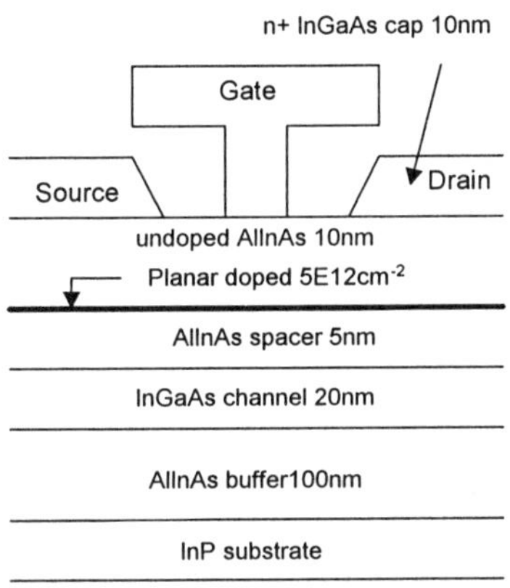

Figure 3-20. Cross-section of LM HEMT device.

The measured drain current versus drain voltage characteristic of a 0.25 μm gate length LM HEMT[32] is shown in Fig. 3-21.

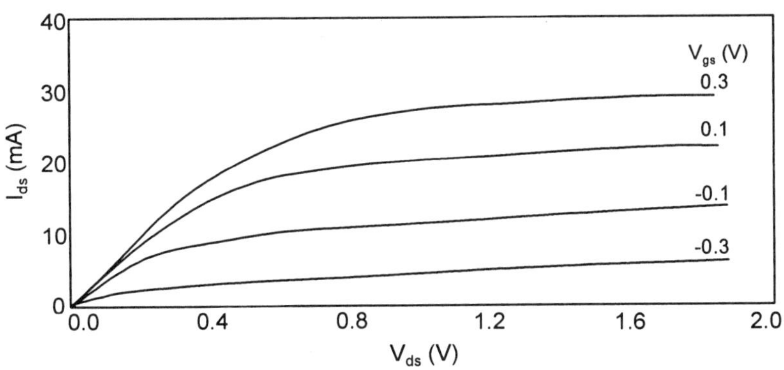

Figure 3-21. Measured LM HEMT (0.25 μm gate length) I - V characteristic (source Duh et al[32], © 2004 IEEE).

A typical drain current versus gate voltage for a 100 μm HEMT device (0.15 μm gate length), from Angelov et al[29], is shown in Fig. 3-22. The transconductance trace for the same device is outlined in Fig. 3-23. The associated figures of merit for this LM HEMT are:

$$I_{max} = 320 \text{ mA/mm}; \; g_m = 370 \text{ mS/mm}$$

These figures of merit are very similar to those of the pHEMT device.

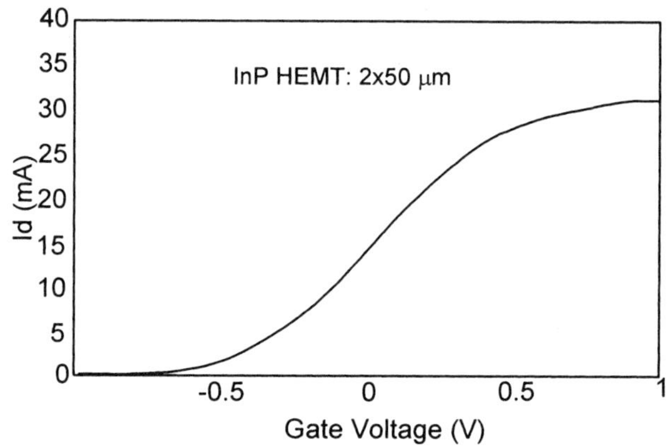

Figure 3-22. Typical InP HEMT (100 μm, 0.15 μm gate length, V_{DS} =2V) I - V characteristic (source Angelov et al[29], © 2004 IEEE).

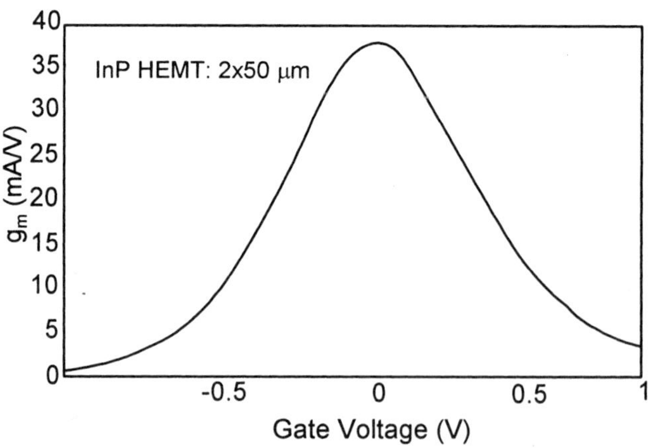

Figure 3-23. Typical InP HEMT (100 μm, 0.15 μm gate length, V_{DS} =2V) g_m - V characteristic (source Angelov et al[29], © 2004 IEEE).

2.4.4 Metamorphic HEMT

The cross section of a typical metamorphic HEMT (MM HEMT) is shown in Fig. 3-24. This MM HEMT is a device that exploits the channel characteristics of the LM HEMT, but with the added advantage that GaAs rather than InP is used as the substrate material. The use of GaAs has a number of advantages including lower cost, ease of handling (less brittle) and the availability of larger diameter wafer material. Despite the fact that the GaAs substrate is not lattice matched to the channel material, an intervening structure of many layers of InGaAs with varying mole fractions provides a graded interface in which the threading dislocations are suppressed and high quality active layers are achieved[34].

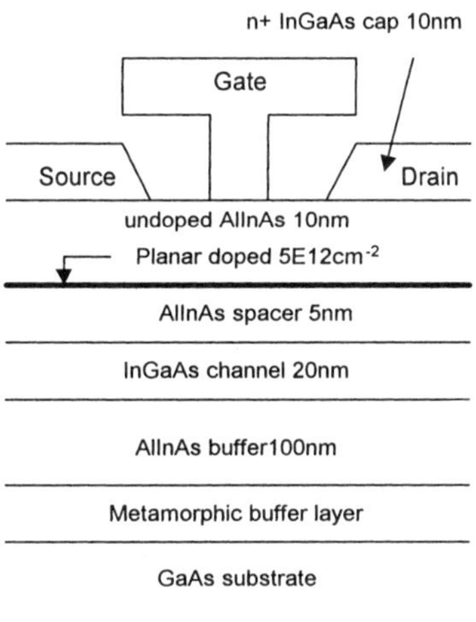

Figure 3-24. Cross-section of MM HEMT device.

In this HEMT structure, high quality channels can be realised for a large InAs mole fraction range. This mole fraction has a significant impact on the material properties. An increased InAs mole fraction leads not only to increased low field mobility (desirable) but also to a reduced conduction band discontinuity (undesirable due to increased leakage currents and reduced breakdown voltage). As a consequence, the optimum mole fraction from a device performance perspective can be dependent on the intended application.

This MM HEMT device is distinguished from its predecessors by the alloy composition of the material structure, which can now be chosen arbitrarily without concern regarding lattice mismatch to the underlying substrate material[21]. Of course, this luxury comes at the cost of complexity, which demands more complicated epitaxial growth processing.

Picogiga[35] became the first GaAs wafer manufacturer to offer epitaxial GaAs wafers designed to facilitate MM HEMT circuit fabrication.

The Fraunhofer Institute (Erlangen, Germany) offers a 0.15 μm MM HEMT process on a foundry basis. Such a process offers the potential for promising transceiver type developments at 100 GHz and beyond.

2.5 FET Equivalent Circuit

It is appropriate at this stage to introduce a very simple equivalent circuit for the MESFET. This circuit will facilitate the derivation of useful expressions for current and power gain for the device. These expressions provide insight into the relationship between the device parameters and its corresponding high frequency performance. A basic equivalent circuit is shown in Fig. 3-25. For fixed component values, this model is valid at a single bias point. Such a model is also called a linear model. It is reasonable to utilise such a small signal model to represent the device for small AC excursions about that bias point. More complex large signal models, which are valid over a range of bias and drive conditions, can also be developed. These are non-linear models, whose element values are not fixed, but are functions of the instantaneous bias conditions applied to the device. A similar equivalent circuit is also appropriate for the more sophisticated FET variants discussed above.

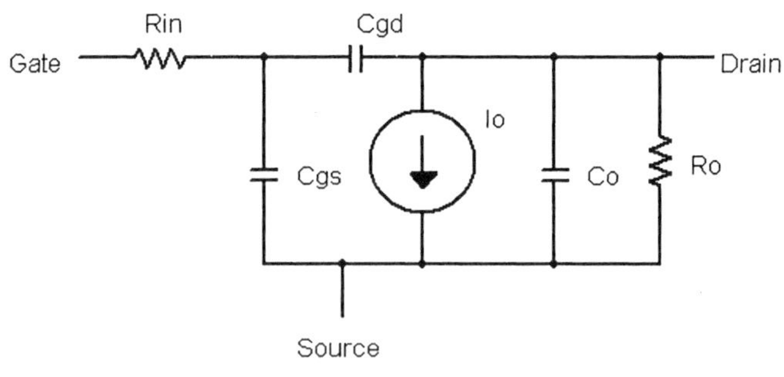

Figure 3-25. GaAs MESFET equivalent circuit.

The key elements of this equivalent circuit are the input resistance to the gate, R_{in}, and the gate source capacitance on the input side. On the drain side of the device (usually the output), the main components are the current source, whose current is related by the transconductance to the voltage across the gate source capacitance, and the output resistance and capacitance. Generally speaking, for good high frequency performance, a large value for R_0 and a small value for C_0 are required so that the output signal is not excessively shunted in the device itself. Small signal equivalent circuit models for transistors are derived from small signal measurements.

A number of extraction methods for the above, or for similar equivalent circuit model formulations, are commonly applied, but these are beyond the scope of this work. Divekar[36] is suggested to the interested reader. Large signal model development is a more complex effort. This requires extensive small signal characterisation of a device at a range of bias points (essentially covering the range of instantaneous bias conditions for which the resulting model is intended to be valid) in conjunction with a set of DC measurements.

Over the years, many large signal MESFET model formulations have been published[37–39]. These models have been successfully applied in a large number of high frequency designs. As CAE tools have evolved and harmonic balance simulation engines have become more sophisticated, circuit designers are demanding models that provide accurate predictions of more involved circuit responses. For example, early models would typically have been used to study amplifier gain responses and matching conditions, whereas modern models are extensively used to simulate intermodulation characteristics, noise figure, circuit response temperature dependence and other parameters for which the older model formulations would not have been suited. The Angelov model[29] is one of the more sophisticated formulations and is applicable to MESFETs, and also to more exotic FET device types including HEMT, pHEMT, LM HEMT and MM HEMT.

Regardless of how the equivalent circuit parameter values are derived, most FET models take the form of Fig. 3-25, with some additional extrinsic lumped elements being added. These additional items represent the parasitics involved going from the internal intrinsic FET device to the outside world, and commonly include parasitic pad capacitances and lead inductances.

In order to assess the likely performance of a given device in a circuit design, values for the elements of the equivalent circuit (either the model above, or perhaps an alternative but similar circuit representation) are required.

A number of the standard models have been built into the high frequency circuit simulator CAE packages available today. Some of the simulators also contain proprietary models developed specifically for those tools.

2.5.1 Linear MESFET Models

Linear models are appropriate for small signal circuit designs. These models essentially provide a best-fit to the actual device characteristics at a given bias point. As the applied signal levels are small, it is assumed that the instantaneous bias point does not vary significantly over time, and that the best-fit model is *time invariant*. In this type of model, the equivalent circuit generally consists of lumped elements with fixed values. The circuit simulators then have the straightforward task of embedding the lumped element circuit in the overall circuit, in the place of the device, and applying conventional circuit analysis techniques to determine the circuit response.

2.5.2 Non-Linear MESFET Models

Non-linear models are appropriate in designs where non-linear effects are likely to occur. In this instance, due to the large levels of some of the applied signals, the instantaneous bias point of the device varies with time. In order to model the circuit response accurately, the model must be capable of modelling the bias dependent nature of the device characteristics. In other words, as the applied signals vary with time, and the effective bias point shifts, the model must re-generate lumped element equivalent circuits as a function of the changing bias conditions.

2.6 Fundamental FET Circuit Relationships

Some important relationships for the model in Fig. 3-25 are developed below. The AC current source is dependent on the input voltage, and this relationship is represented by:

$$I_o = g_m v_{gs}$$

where g_m is the transconductance. This follows directly from the definition of g_m which is:

$$g_m = \frac{\partial i_{ds}}{\partial v_{gs}}$$

Now, consider the short circuit current gain, h_{21}, (i.e. the current gain with a short circuit connected across the load which essentially ensures that the full current I_o flows through that load).

$$h_{21} = \frac{I_{out}}{I_{in}}$$

$$I_{out} = I_o = g_m v_{gs}$$

$$I_{in} = \frac{V_{in}}{R_{in} + 1/j\omega C_{gs}}$$

$$v_{gs} = I_{in}/j\omega C_{gs}$$

$$\therefore \quad h_{21} = \frac{g_m}{j\omega C_{gs}}$$

The transition frequency, f_t, is defined as the frequency where h_{21} becomes equal to unity. From the above expression for h_{21}, it is immediately clear that f_t is given by:

$$f_t = \frac{g_m}{2\pi C_{gs}}$$

This shows clearly that high frequency operation of a MESFET transistor requires a high transconductance and a low gate-source capacitance.

Now, consider the voltage gain of a MESFET. It is well known that the gain is maximised when the input and output are both conjugate-matched. The impedance seen looking into both ports of the device can be expressed as:

$$Z_{in1} = R_{in} + j\omega C_{gs}$$

$$Z_{in2} = R_o \| (1/j\omega C_o)$$

For maximum power transfer to occur, the source and load impedances need to be the conjugates of the above. As a consequence, the reactive components are cancelled. Applying the conventional result for the power available from a source, and assuming that all voltages are RMS values, the power available at the *input* port of the MESFET is

$$P_{avail,in} = \frac{|V_{in}|^2}{R_{in}}$$

Now, consider the gain of the MESFET itself, under the condition that the output is conjugately matched. This means that the effective resistance seen in parallel with the current source is ½ R_o. Thus, the output voltage is given by

$$V_o = -g_m v_{gs} \frac{R_o}{2}$$

But, it is clear that

$$v_{gs} = V_{in} \cdot \frac{1/j\omega C_{gs}}{R_{in} + 1/j\omega C_{gs}} = \frac{V_{in}}{1 + j\omega C_{gs} R_{in}},$$

and hence the output power in the matched resistance is

$$P_{L(\max)} = \frac{|V_o|^2}{R_o} = \frac{1}{R_o}(g_m R_o/2)^2 \frac{|V_{in}|^2}{1 + (\omega C_{gs} R_{in})^2}$$

Thus, the maximum power gain is

$$G = \frac{P_{L(\max)}}{P_{avail,in}} - \frac{(g_m)^2 R_o R_{in}}{4[1 + (\omega C_{gs} R_{in})^2]}$$

Typically,

$$\omega C_{gs} R_{in} >> 1$$

and the available power gain reduces to

$$G = \frac{(g_m)^2 R_o}{4(\omega C_{gs})^2 R_{in}}$$

This gain is also known as the unilateral power gain, U, and can be reduced to

$$U = \frac{1}{4f^2}\left(\frac{g_m}{2\pi C_{gs}}\right)^2 \frac{R_o}{R_{in}}$$

U is also defined as

$$U = \left(\frac{f_{\max}}{f}\right)^2,$$

where f_{max} denotes the maximum frequency at which the device is capable of oscillating. From the above, it is clear that

$$f_{\max} = \frac{f_t}{2}\sqrt{\frac{R_o}{R_{in}}}$$

Thus a high frequency MESFET requires not alone a high f_t, but also a high output resistance (or low conductance) and a low input resistance. Intuitively, this makes sense - the low output conductance prevents the load from being shunted by the device itself, and the low input resistance ensures that the gate-source capacitance sees the majority of the applied signal, and as a result the output current source is strongly correlated with the applied signal level via the transconductance. Both f_t and f_{max} are commonly cited as figures of merit for a high frequency device.

2.7 GaAs Hetero-Junction Bipolar Transistor

The hetero-junction bipolar transistor (HBT) is a more sophisticated variant of the conventional bipolar transistor. A schematic cross section of a HBT structure is presented in Fig. 3-26.

Among the issues with the use of a conventional (Si) bipolar transistor at high frequencies are the high base resistance and the large parasitic capacitances. First proposed by Kroemer[40] in 1957, the HBT device offers superior performance due to lower values of the parasitic resistance and capacitance. The device did not become a reality until much more recently when sufficiently advanced fabrication processing techniques were developed. The HBT makes use of a hetero-junction between the emitter and base. This junction, with a wider band-gap material in the emitter (typically

AlGaAs or more recently InGaP) results in an offset in the valence band such that hole injection from the base to the emitter is retarded while, at the same time, the electron injection into the base is enhanced. As a consequence, it is possible to increase the base doping while at the same time maintaining a good β. The increased doping leads to a lower base resistance. Similarly, the emitter doping can be reduced, thereby resulting in a lower base-emitter capacitance. The characteristics of GaAs (and other III-V compound semiconductors) ensure that the transit times for HBTs are considerably shorter than they are for equivalent geometry Si based bipolar transistors.

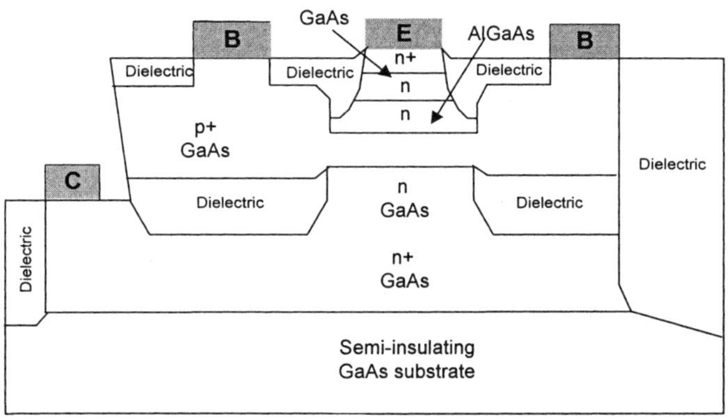

Figure 3-26. Cross-section of a hetero-junction bipolar transistor on GaAs.

In contrast to a FET type device, where the critical dimensions are defined by lithography, the key distances are established by epitaxial growth in a HBT. This makes possible a high f_t with modest fabrication processing requirements. The full emitter area can carry current, thereby yielding high current capacity and high g_m for the HBT. The HBT device also tends to have a lower output conductance than a FET as the active region is better shielded from the output voltage. Combined with the higher g_m, this leads to a higher voltage gain. The HBT device is well shielded from trap effects, which tend to degrade the *1/f* noise of FETs; consequently, the HBT is more suited to oscillator type applications. Most HBTs are fabricated by MOCVD, mainly due to economics and the availability of very high doping levels for the base using carbon.

The material layer structure of a typical GaAs/AlGaAs HBT is shown in Fig. 3-27. Noteworthy features include the graded composition each side of the $Al_{0.3}Ga_{0.7}As$ emitter, very high p-doping using carbon in the thin base, and a heavily doped sub-collector. The use of carbon has had a large impact on HBT technology, since the previous p-doping materials diffused during growth leading to non-abrupt doping profiles and non-optimum junction location[41, 42].

The corresponding layer structure for an InGaP based HBT is presented in Fig. 3-28.

Much work on the use of C-doping for the base region of HBTs has been done by a University of Illinois group[43]. Today, the usefulness and stability of this dopant are no longer in doubt. The use of InGaP in place of AlGaAs in the emitter avoids the use of Al and enables simple selective etching of both GaAs and InGaP. The use of InGaP gives similar or improved performance compared to AlGaAs, while simplifying the device fabrication and aiding reproducibility. The InP/InGaAs system has also been used to fabricate HBTs, but this system is not yet as advanced technologically as the GaAs based materials. GaAs HBT circuits with good mm-wave performance have been reported recently, including a 60 GHz GaAs/InGaP VCO[44].

Due to the vertical nature of the typical HBT, the die or chip can be very small. This allows circuit designers to pack many HBT circuits onto a wafer. This high die count, combined with the simpler, higher yield HBT process, enables companies to generate many circuits and more revenue from each processed wafer, while lowering circuit prices to their customers.

In general, for high frequency applications, the HBT device is considered a viable option for oscillator and power amplifier requirements. For the target converter application areas considered in this work, the FET type devices were considered more suitable as the device technology and the HBT will not be considered further.

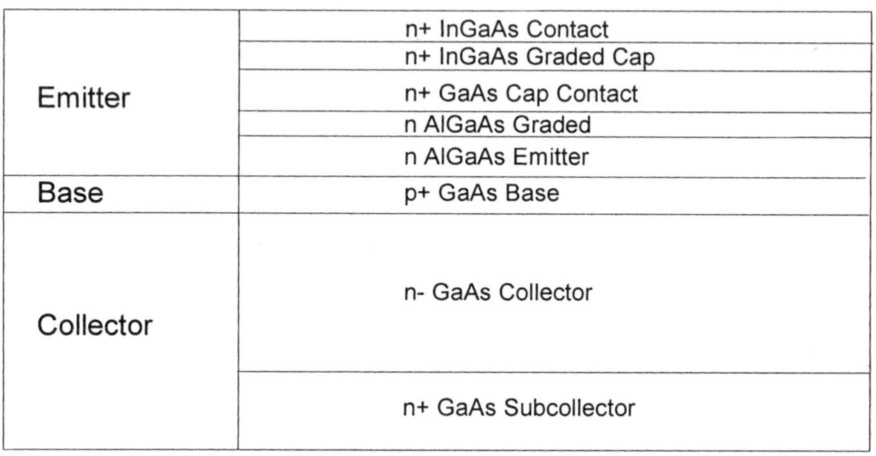

Figure 3-27. Material layer structure of a typical GaAs/AlGaAs HBT.

Emitter	n+ InGaAs Contact
	n+ InGaAs Graded Cap
	n+ GaAs Cap Contact
	n InGaP Emitter
Base	p+ GaAs Base
Collector	n- GaAs Collector
	n+ GaAs Subcollector

Figure 3-28. Material layer structure of a typical GaAs/InGaP HBT.

2.8 Silicon based High Frequency Devices and Circuits

Traditionally, Si integrated circuits such as those found in consumer electronics goods and many other applications have used Metal Oxide Semiconductor Field Effect Transistors (MOSFETs) and Bipolar Junction Transistors (BJTs), but neither of these transistors operate above a few GHz because of the material properties of Si. In recent years, a new material system based on epitaxial SiGe layers on Si has been developed which permits the fabrication of HBTs and HEMTs that operate into the mm-wave frequency spectrum and thus make Si a viable material for microwave circuit applications. Many research groups are pursuing studies on material growth methods and properties, device fabrication and performance, circuit design, and commercial applications[11].

There are several reasons that this research work is being carried out. Firstly, it is generally accepted that high frequency Si circuits will cost less to fabricate than the traditionally used GaAs and InP based microwave integrated circuits because of the larger Si commercial market and the associated competition, and of course the more mature fabrication technology associated with Si. Secondly, lower power levels may be used which should accelerate the growth of hand-held communication devices. Thirdly, and most importantly, integration of RF SiGe technology with digital CMOS circuits would enable single chip communication systems. These single chip systems would have smaller packages and less interconnects which would result in lower cost and higher reliability. This last reason is the Holy Grail in terms of high frequency circuit development for commercial applications.

Unfortunately, good transistors are not enough to fabricate good high frequency circuits; low loss transmission media and high Q passive components are also required. Since the resistivity of common CMOS grade silicon wafers is in the range of 1-20 Ω*-cm*, circuit elements placed directly on the Si have high loss. Various researchers have studied a number of possible solutions to this issue. The easiest approach is to use commercially available high resistivity Si wafers since all of the circuit elements may be implemented in the same way as they are on GaAs, InP, or ceramic substrates. This has been demonstrated for coplanar waveguide and coplanar stripline transmission lines, distributed and lumped circuit elements, and antennas. The issue here of course is that the same substrate would not be ideal for a CMOS circuit, and hence the end goal of integrating the high frequency and conventional electronic circuits on a single substrate is not achieved. Another technique that has been demonstrated is to use thin layers of polyimide to fabricate transmission lines that are completely shielded from the Si wafer. Other relevant fields of research include Silicon-on-Sapphire (SOS)[45] and Silicon-on-Insulator (SOI)[46]. Finally, MEMS fabrication techniques may be used to remove the Si around the transmission lines and circuit components. This MEMS technology has been pioneered, among others, by Professor L. P. Katehi at the University of Michigan[47]. Whichever of these, or possibly other, techniques is deemed the optimum to achieve good high frequency circuit performance on a silicon substrate, it is clear that the fabrication process involved is likely to be a significant departure from the conventional CMOS processes as they exist today.

3. THE FUTURE...

The future in terms of high frequency electronic devices will follow many paths. For certain, the development of high frequency devices on Si substrates will continue with the huge associated potential integration benefits. The obstacles to progress, such as the realisation of quality low loss passive elements, will no doubt be overcome, and system on chip solutions will become a reality.

However, III-V compound semiconductors will continue to improve, and the evolution of associated transistor devices will ensure that these transistors will always have more capability than their Si counterparts. These super-high frequency transistors will remain a driver for high frequency application developments, further extending the frequency range for which commercial exploitation is feasible.

Chapter 4

HIGH-VOLUME mm-WAVE CIRCUITS

1. THE CHALLENGE

The challenge associated with this work is to help to facilitate the true commercialisation of the mm-wave spectrum using monolithic integrated circuits. This will pave the way for a whole range of new high-volume applications enabled by the broad bandwidths available at mm-waves.

It is self-evident that broadband applications, and especially those with bandwidth requirements of the order of hundreds of MHz and beyond, demand relatively high carrier frequencies, and are particularly suited to the mm-wave region of the spectrum. Applications such as automotive collision avoidance, Multipoint Video Distribution System (MVDS) and some WLAN systems[1, 2] have already demonstrated the potential for mm-wave products in the commercial arena. However, these developments to date have typically made use of waveguide technology to varying degrees. In the case of automotive radar systems, the waveguide content is significant and generally includes at least a Gunn oscillator. In MVDS transceiver units, the waveguide element has been reduced to an extent with the recent availability of mm-wave MMIC chipsets[48]. However, it has been the experience of the authors that such chipsets can be exceedingly difficult to work with. In particular, the direct cascading of building blocks can be troublesome due to non-ideal (i.e. non-50 Ω) port impedances being realised with some parts.

In order to make possible the true exploitation of the mm-wave region of the spectrum, using monolithic technology, for low cost communication systems, the key requirements involve building blocks suitable for integration in a mm-wave transceiver. These will make possible the replacement of a major portion of the traditionally expensive waveguide-based material with parts that are equivalent from a functional perspective but which are cheaper, lighter, smaller, and generally more reliable. To make this a reality, suitable design methodologies are required. It is insufficient that good performance is achieved with a mm-wave design – the performance must be repeatable, and must be realisable with high frequency processes that are not necessarily state of the art but which are of course suitable for mm-wave developments. A good fundamental understanding of circuit performance is key, and sound design techniques are critical to success. Circuit designs with broadband performance of course expand the range of applications for which a given MMIC design may be suited.

In this chapter, transceiver topologies suited to mm-wave developments are outlined, and the building block requirements discussed. These building blocks will form the core of this work both from a fundamental understanding and a practical implementation perspective, and will be considered in detail in subsequent chapters. The viability of integrating the building block functions into a single die transceiver will also be discussed in a later chapter and conclusions will be drawn about future developments in this regard.

2. TRANSCEIVER CONFIGURATIONS FOR mm-WAVES

A mm-wave transceiver is a sub-system which provides both up and down-conversion functionality for a mm-wave communications system. The transceiver supports both the transmit and receive functions of the system. In its usual and most basic manifestation, the transceiver consists of a pair of mixers, and a local oscillator, possibly shared between the mixers.

2.1 Existing Solutions

Various transceiver implementations, either used in commercial products or reported in the literature, are described in this section.

The conventional architecture is depicted in Fig. 4-1. This depicts a pair of mixers driven by a common LO source via a splitter. Such a configuration would typically be implemented in waveguide technology using Schottky diode mixers, with the LO consisting of a Gunn oscillator.

The dotted box encompasses what would generally be considered the transceiver. The antennas would typically have waveguide interfaces which mate directly with the RF waveguide ports of the mixers. The IF interfaces would usually be coaxial, possibly SMA type depending on the frequency range involved.

In some waveguide-based transceivers, a single mixer is used to provide both the up and down-conversion functionality at the same time. This exploits the non-linear characteristic of the Schottky diode which ensures that efficient up and down-conversion can be simultaneously achieved in the same mixing element. Such a configuration is presented in Fig. 4-2.

In this transceiver, an optional isolator is also included between the LO source (again generally a Gunn oscillator) and the mixer. This isolator could also be included between the oscillator and the power splitter in Fig. 4-1. The approach depicted in Fig. 4-2 has the clear advantage over the earlier topology that only a single antenna and mixer are required. However, there is additional complexity at the IF port where the receive and transmit signals must be separated; this is typically achieved by means of a diplexer and requires different receive and transmit intermediate frequencies. For typical low IFs, a low-cost diplexer can be implemented in a hybrid of planar and surface mount technology.

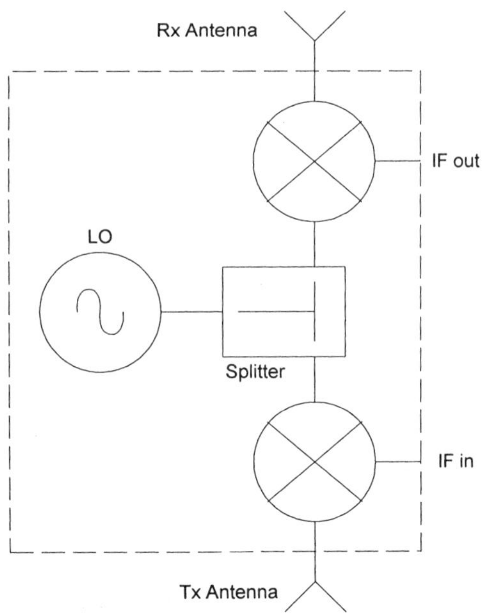

Figure 4-1. Conventional mm-wave transceiver block diagram.

For broadband communication systems, the purity of the LO sources is important. This is particularly the case with high-speed digital transmission where complex modulation schemes are often employed to increase the capacity of the communication channel. In such systems, low phase noise sources are usually required. Achieving good low phase noise characteristics with a mm-wave Gunn oscillator is not trivial. Generally, it requires phase locking to a higher quality lower frequency signal. Suitable phase locking hardware is often available, but it is expensive.

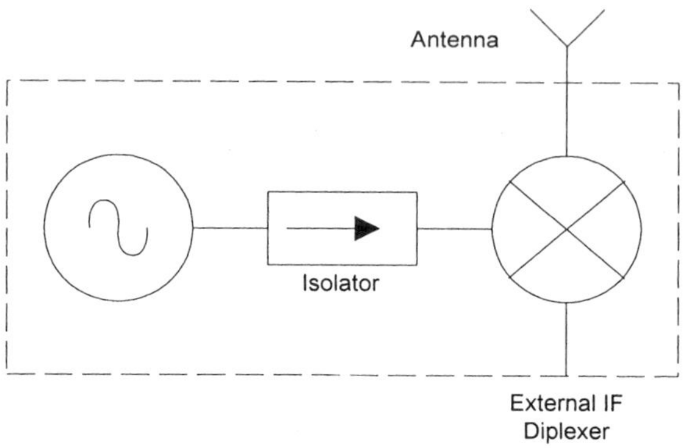

Figure 4-2. Single mixer based waveguide transceiver block diagram.

An alternative approach to the direct generation of mm-wave power with low phase noise is to generate a high-quality lower frequency signal and employ a frequency multiplier to deliver the mm-wave output at the desired frequency. This approach is commonly adopted in mm-wave transceivers. Depending on the phase noise requirement for the mm-wave LO signal, the quality of the lower frequency input source is determined. An example of a transceiver incorporating a frequency multiplier is shown in Fig. 4-3.

The only respect in which this approach differs from that presented in Fig. 4-1 is in the LO provision.

All of the transceiver architectures discussed thus far have evolved from waveguide based origins. In this respect, they do not contain any amplifiers. In recent years, some mm-wave transceiver manufacturers have enhanced the performance capability of their products by incorporating mm-wave amplifiers on the receive, transmit, or sometimes both arms. These amplifiers, which are often based on the hybrid integration of high frequency transistors on typically alumina carriers, tend to be very expensive.

2.2 Monolithic Solution

The transceiver topology in Fig. 4-3 lends itself to monolithic implementation. It assumes a common LO shared between the up and down-converting arms. As mm-wave amplifiers are now realisable on high frequency processes using monolithic technology, the monolithic transceiver topology can also incorporate a low noise amplifier on the receiver chain input, and/or a power amplifier unit on the transmit arm. In this way, the functionality offered at great cost by introducing expensive hybrid amplifiers into waveguide based transceivers, can be made available at much lower cost in a monolithic equivalent product. The inclusion of amplifiers enhances the performance of a monolithic transceiver. This enhancement may manifest itself on the receive side as a reduced system noise figure and improved sensitivity; on the transmit side, the transmit level and the associated linearity can be improved. A potential monolithic topology is shown in Fig. 4-4. The building blocks pertinent to the development of such a transceiver will be discussed in the remainder of this work.

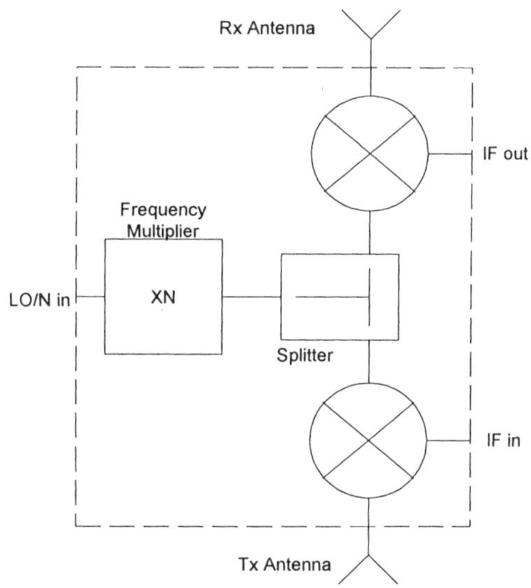

Figure 4-3. Frequency multiplier based mm-wave transceiver block diagram.

3. INTEGRATED MONOLITHIC TRANSCEIVER CONSIDERATIONS

The monolithic transceiver makes possible several benefits, including cost, weight and size reduction and enhanced reliability. The inevitable disadvantage, in comparison to the waveguide based topology in Fig. 4-2, is the requirement for two antennas. However, with the advances in planar antenna technology, this is becoming a small price to pay. Moreover, many waveguide transceivers already employ two antennas (Figs. 4-1 and 4-3) and in that context, the monolithic approach does not introduce any additional complexity at all.

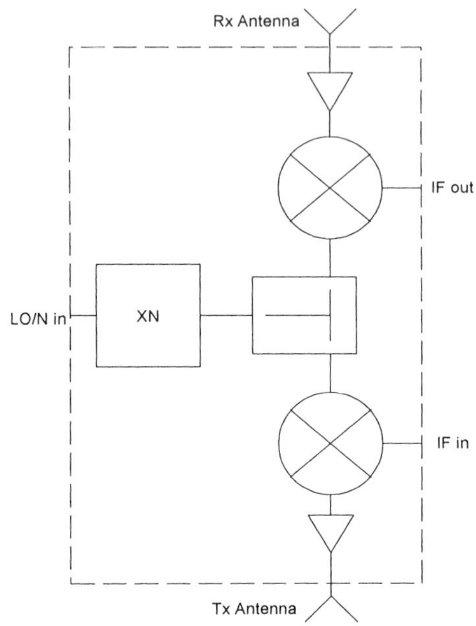

Figure 4-4. mm-wave monolithic transceiver topology.

The natural extension to a monolithic implementation of the building blocks in a transceiver is to integrate some of those building blocks into a single die. This leads to a lower assembly cost due to a reduced bond-wire count. The package size can also be reduced. It is also reasonable to assume that the reduced number of bonds will improve the reliability of the assembly. The design of an integrated transceiver is not performance-constrained by the internal bond-wires and their associated impact on, for example, realisable bandwidth. This makes the design of the building blocks, and the interface between them, more straightforward.

A key aspect of an integrated monolithic transceiver that needs to be considered is yield. The tradeoff that must be assessed is the likely yield impact of adopting the fully integrated approach instead of the individual MMIC building blocks. A simple model for this comparison is now presented.

Consider a very simple model where there are p defects per wafer of size W units. Additionally, assume the individual circuit area is c units, and assume there are N different circuit types which are combined to complete the monolithic transceiver. In order to simplify the analysis, assume that the different circuit types have the same area and that they are all fabricated on the same wafer – of course this is not the typical case in reality but is a reasonable assumption to make in order to gain a useful insight.

The number M of each type of circuit on a wafer is given by

$$M = W/cN$$

The average number of defects per individual circuit then becomes

$$d = p/MN = pc/W$$

Assuming $d < 1$, the corresponding probability that each individual circuit is defect-free is

$$Y = 1 - d$$

Now, consider two distinct approaches to implementing the monolithic combination of the building blocks: firstly, using the individual building blocks and packaging them together (i.e. a multi-chip assembly) – denote this by *Individuals*; secondly, integrating the individual building blocks into a single MMIC – denote by *Integrated.*

In the first case, *Individuals*, the probability that the combination of N individual MMICs packaged together is defect-free is given by

$$Y_{indiv} = (1 - d)^N$$

On the other hand, in the case of *Integrated*, the likelihood that this single chip is free of defects is given by

$$Y_{\text{integ}} = (1 - Nd)$$

It is now instructive to compare the corresponding failure rate characteristics, where failure rate is defined as

$$\text{Failure Rate (\%)} = 100\ (1 - Y)$$

Such a comparison, for the model developed above, is depicted in Fig. 4-5 for values of d ranging from 0 to 0.1, and N values of 1 and 5. The $N = 1$ case is the trivial scenario where only one building block is used in the fully integrated circuit and the failure rates are necessarily identical in this case. For the $N = 5$ case, it can be seen that as d increases, the *Individuals* failure rate is lower than that associated with *Integrated*. The clear conclusion one can draw from this very simple analysis is that as the number of different circuits in the integrated solution increases, the yield of the individual circuits packaged together becomes increasingly higher than the yield associated with the corresponding single integrated MMIC solution. In other words, the integrated MMIC yield becomes increasingly unattractive, as the design becomes more complex.

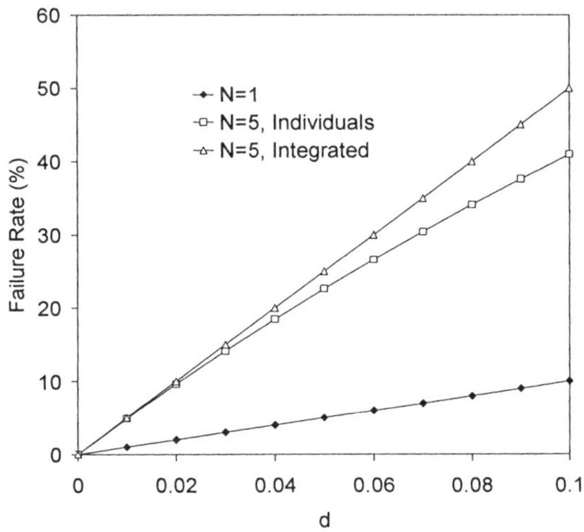

Figure 4-5. Failure rate versus d for Individuals and Integrated transceiver options.

The challenge then becomes how to achieve an overall yield for a fully integrated transceiver design such that the inevitable yield drop associated with higher levels of integration is insufficient to offset the cost and other benefits associated with the more integrated approach. There are two aspects associated with the realisation of such a high yield. Firstly, the fabrication process must be very consistent, and considerable effort has been and continues to be expended by many researchers in this regard. Secondly, the design itself must be optimised with yield in mind. This concept is often referred to as design for manufacturability, and an absolutely key component of this is a strong fundamental understanding of the functionality of the building blocks used in the design such that optimised robust designs can be developed. The improvement of this fundamental understanding forms a large part of the original work presented in this work. By facilitating an optimum yield realisation, and thereby aiding the viability of fully integrated monolithic transceivers, this book adds to the body of work that brings the widespread use of GaAs MMIC technology for high volume mm-wave applications closer to being a reality.

4. THE GOAL

In light of the clear benefits associated with the successful development of mm-wave MMIC components suitable for use in transceiver systems, the goal of this work has involved the study of the key building blocks involved. This study comprises circuit analysis at a fundamental level where possible, backed up by a substantial practical element with mm-wave MMIC design, fabrication and test. The combination of analysis and verification led to insightful design methodologies which, if employed, ought to improve the quality of high frequency circuit design.

This building block study has been carried out in the context of two distinct application areas. Firstly, the 57 GHz short-hop radio transceiver application area, and secondly the 40 GHz MVDS market. The former is a loosely regulated radio link application that is widely used for security camera type applications. It exploits the atmospheric attenuation characteristics near 60 GHz which ensure that the link is inherently short range and secure (signals are unlikely to be picked up except by those receivers for which the link is intended). This is likely to be a medium volume market, and the requirements on the quality of the externally applied low frequency LO are relatively relaxed. The MVDS market is an interactive broadband service aimed at distributing data and voice to the home. It was originally conceived as a means of broadcasting analogue multi-channel television stations to homes on a regional basis, but has more recently evolved into a fully interactive data service (of course including digital television). This is likely to be a very large market and the broadband nature of the signals involved places more stringent demands on the quality of the external low frequency LO in this case. In fact, this requirement for a very pure LO in the case of MVDS in turn demands that a low frequency high quality source be used with a frequency multiplier as described earlier in this chapter. In the case of the 57 GHz radio links, direct generation at mm-waves is not precluded. However for a monolithic implementation, the frequency multiplier again is the approach of choice. The building blocks investigated during the course of this work include a frequency multiplier, mixers and amplifiers for the 57 GHz radio link market, and a frequency multiplier and an amplifier suitable for MVDS requirements. The linkages between the building block MMICs designed and evaluated in this work (Chapter 1, Fig. 1-1), and the typical monolithic transceiver in Fig. 4-4, is illustrated in Fig. 4-6. The dotted links in Fig. 4-6 are associated with the amplification elements in the transmit amplifier and in the active splitter.

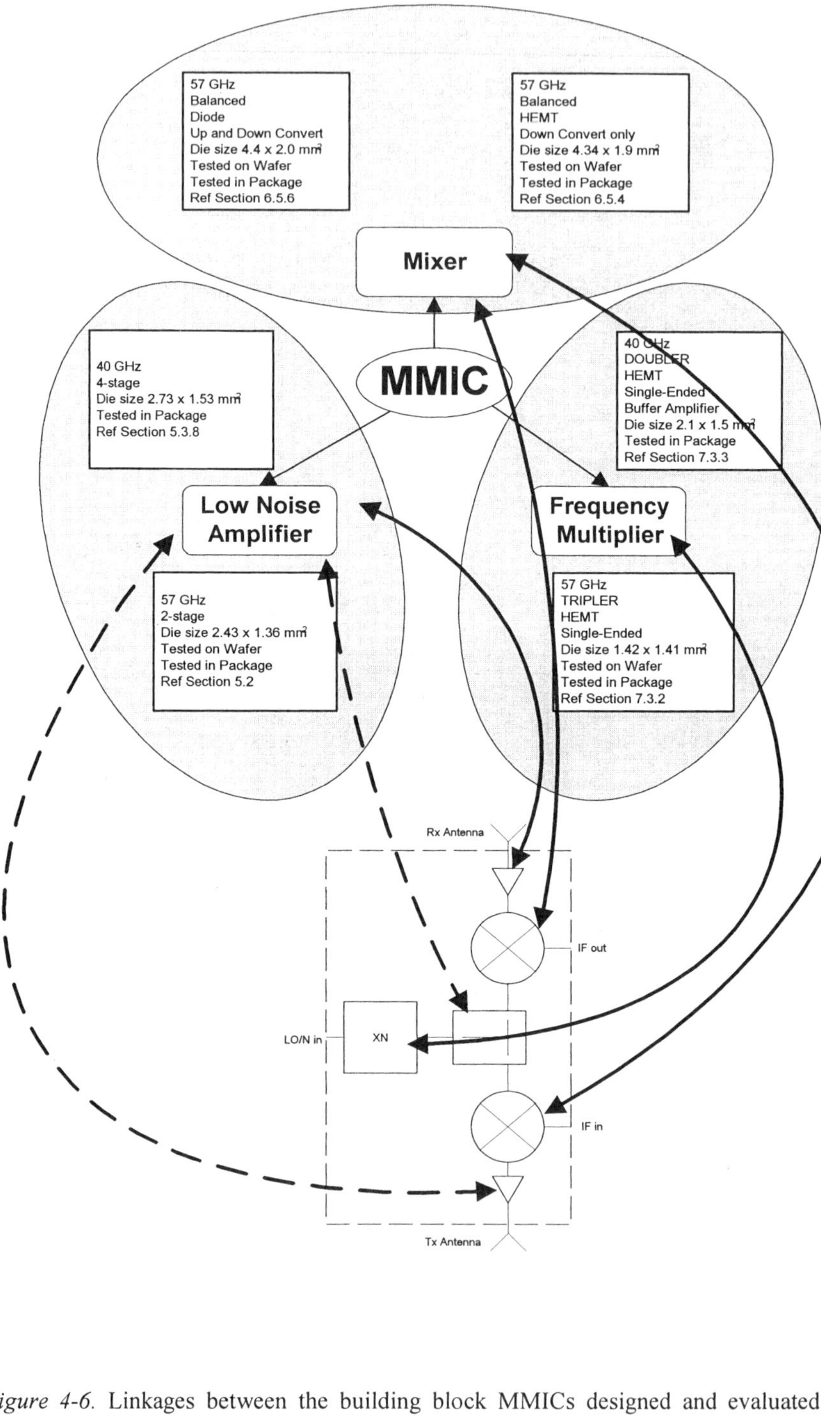

Figure 4-6. Linkages between the building block MMICs designed and evaluated in this work.

Neither of these building blocks was developed specifically by the authors (they were designed by a colleague of the authors[49]), but the design techniques outlined in Chapter 5 are also directly applicable to those circuit developments.

A secondary goal of this work was to assess the suitability of the building blocks for further integration with a view to exploiting some of the potential benefits of this approach, as outlined earlier. This study of a “transceiver on a chip” solution for the 57 GHz application includes the fabrication and test of a practical realisation on GaAs. The potential of such an approach for high volume manufacturing is discussed, and the changes needed to facilitate the volume manufacturability requirements are considered.

Chapter 5

LOW NOISE mm-WAVE AMPLIFIERS

1. INTRODUCTION

A critical component in many receiver type applications is the low noise amplifier (LNA). This is a high sensitivity circuit which amplifies an incoming low-level signal while degrading the associated signal to noise ratio to a minimal extent. The gain of the amplifier reduces the contribution of subsequent receiver components to the overall system noise figure. The sensitivity of a modern receiver is largely driven by the performance of its front-end LNA.

Until recently, high performance mm-wave LNAs were not realisable, and typical mm-wave (and higher frequency) receiver architectures had a down-converting mixer as the front-end component after the antenna. As a consequence, the system noise figure was constrained by the noise performance (typically 6-8 dB) of what was typically a Schottky diode based waveguide mixer. Moreover, due to the conversion loss of this mixer (about 6 dB), the noise figure of subsequent elements in the receive chain also contributed significantly to the system noise. With the relatively recent emergence of GaAs fabrication processes with adequate high frequency capability, mm-wave LNAs have become a reality[50]. This in turn makes possible mm-wave receivers with enhanced noise performance, which makes possible the development of more sophisticated high frequency systems than would be possible with the more conventional front-end approach. As high volume mm-wave applications such as MVDS and WLAN emerge, there is a growing demand for robust high performance LNA designs operating at high frequencies.

Design techniques and guidelines for low noise amplifiers are widely published[51]. Typically, these involve the presentation of the optimum input reflection coefficient (or impedance), Γ_{opt}, to the active device. When this impedance is presented, the device noise figure is minimised. This reflection coefficient is a function of the device bias and the source terminal (or emitter for bipolar transistor) termination. Typically, for a FET type device, it is found that the absolute minimum noise figure at lower frequencies is obtained for a bias current of approximately 10% of I_{dss}. Under this condition, the gain is usually not degraded significantly from its typical value at 50% of I_{dss}. The low current may degrade the linearity (and hence dynamic range), but this may not be a major concern for low-level signal inputs.

However, mm-wave LNAs typically require a slightly different emphasis at the fundamental design stage. In particular, due to limited device performance at high frequencies, it is often not a viable option to back off the current to the optimum bias point indicated above, because the associated gain may drop excessively. This chapter addresses the specific requirement of a LNA design methodology suitable for mm-wave frequencies. It is by no means the only design approach, but in the authors' opinions, represents an interesting and novel design methodology that is comprehensible, and has the additional benefit of being eminently suited to straightforward cascading of stages to increase the overall amplifier gain.

The new design methodology is discussed in the context of a MMIC LNA design developed to address requirements in the 30 – 50 GHz frequency range. The primary target application was MVDS but a broad-band design was desirable to make the die suitable for as many application areas as possible. The final MMIC layout contains four stages, but some simulations for a three-stage amplifier studied during the design cycle are also presented in order to emphasise and demonstrate how the response bandwidth can be broadened. The excellent measured performance realised with this design is discussed in detail, particularly in the context of a comparison with simulation.

The chapter opens with a discussion of a 57 GHz LNA circuit that was also designed during the course of this work[49, 52–54], in advance of the new sophisticated approach above being developed. This circuit was designed for fabrication on the GMMT H40 process[3] using conventional techniques, with no significant effort being focussed on making the performance broad-band and robust. The performance realised with this design is highly encouraging, particularly in light of the high frequency range involved relative to the capability of the H40 process. Nevertheless, the merit of this 57 GHz circuit design and the corresponding measured performance fall short of what was achieved with the 30 – 50 GHz circuit, and this provides further affirmation of the new design methodology which comprises the primary topic of this chapter.

2. CONVENTIONAL 57.2 – 58.2 GHz LNA DEVELOPMENT

2.1 Introduction

In this section, design details of a 57 GHz LNA development are presented. The design proceeded along conventional lines, involving extensive use of simulation tools to perform circuit optimisation. The design approach is not overly intuitive, and could reasonably be referred to as a 'blind-faith' methodology. The 57 GHz LNA design focussed primarily on the realisation of performance over a relatively narrow band in the frequency range of interest and realizing broadband performance was not a priority.

This design is interesting in its own right, and presents a useful context for the more sophisticated methodology described in Section 3 of this chapter. From the perspective of the H40 foundry process, it is an entirely novel development, representing the first V-band LNA designed for that process.

The intended application for the circuit is the 57 GHz transceiver requirement discussed previously in Chapter 4. The LNA is required to operate over the 57.2 – 58.2 GHz frequency range with the best possible performance. The primary factors of interest are the noise figure and gain. The linearity requirements are not significant, as the power levels involved in such radio links are quite low.

2.2 57 GHz LNA Design Details

The LNA is a two-stage design based on a 2x30 μm pHEMT device. The two stages are essentially a single-stage design cascaded with itself. Its circuit schematic is presented in Fig. 5-1. The bias networks include on-chip decoupling with appropriate resistors to improve stability. The resistors on the transistor gates are embedded between small and large value decoupling capacitors to ensure that the resistances impact only on the low frequency stability of the circuit and do not compromise the wanted-band noise performance. The large and small value capacitors in the gate bias networks are realised using different dielectric layer structures – the large capacitor utilises a single layer only, whereas the smaller value is achieved by the addition of another layer. Stub matching is employed throughout the circuit and the bias networks comprise part of this matching. Source peaking is provided at the source contacts; this takes the form of an inductance between the source and ground and it helps to move the optimum input impedance for noise (Γ_{opt}) closer to the simultaneous conjugate input match impedance which is desired for maximum gain. However, source peaking also reduces the available gain. As gain is at a premium at the mm-wave frequencies of interest, the design cannot tolerate too much source peaking because the reduced first stage gain results in an increased contribution from subsequent stages to the overall noise figure. The source peaking is realised in the layout as short lengths of transmission line, as distinct from the lumped element inductance approach more commonly used at lower frequencies. The amount of source peaking introduced was optimised in the circuit simulation phase of the development. The final simulated DC bias point was 30% of I_{dss} for both pHEMT devices.

The simulated circuit performance, using the HP-Eesof Libra simulation tool[27], for this amplifier is summarised in Table 5-1.

Versions of the amplifier were designed to suit both the fully integrated transceiver environment (with an input bond-wire only, and a direct 50 ohm load at the output – see Chapter 8), and also the standalone 'packaged' amplifier environment with bond-wires at both ports. These two designs are very similar. A single-stage extraction from one version was also laid out on the mask for diagnostic purposes.

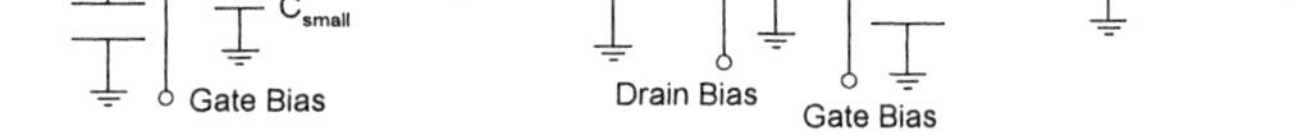

Figure 5-1. Low noise amplifier schematic.

The layout of the standalone two-stage version is shown in Fig. 5-2. RF on-wafer probing is facilitated by the GSG probe pads (see Chapter 2, Section 6.3) at both input and output.

2.3 57 GHz LNA Characterisation

Upon completion of the wafer fabrication processing, on-wafer measurements were attempted on the two-stage designs, using the test facilities of the University of Glasgow (Wiltron 360 Vector Network Analyser with V-band capability), but due to inadequate bias decoupling in the region of the input DC needle probes, oscillations (in the range 10 – 20 MHz) were found to be a considerable difficulty. In fact, it was only possible to measure, on wafer, a stable amplifier response for the single stage version of the LNA. The corresponding maximum value of measured gain was slightly greater than 5 dB near 53 GHz. No test equipment to facilitate on-wafer noise figure measurement at V-band was available. The single-stage amplifier is more immune to oscillation effects than a multi-stage design by virtue of its lower gain and also as a consequence of fewer opportunities for undesired feedback.

Table 5-1. Simulated performance parameters for two-stage 57.2 – 58.2 GHz LNA

Frequency Range	57.2 - 58.2 GHz
Gain	> 10 dB
Noise Figure	< 5 dB

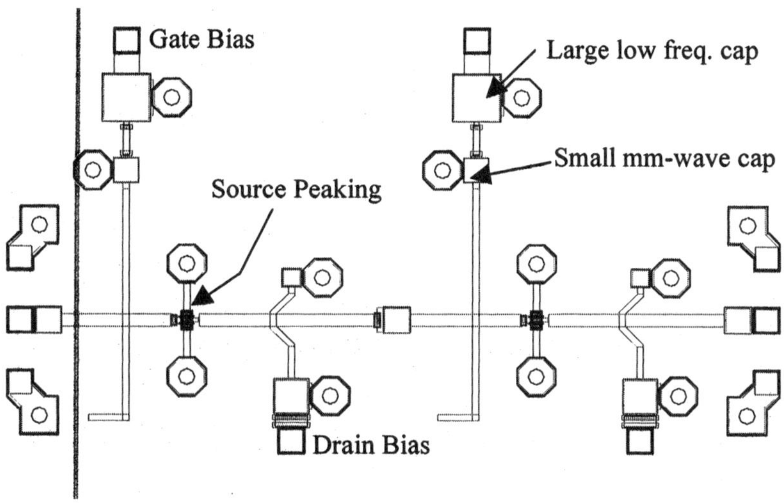

Figure 5-2. Two-stage 57.2 – 58.2 GHz LNA layout.

The measured single-stage amplifier gain versus frequency is compared with the simulated response in Fig. 5-3. It is apparent that the measured performance tallies reasonably well with expectation and this in turn suggests that the amplifier in general is functional. As a consequence it was decided that testing the two-stage amplifier in a suitable package with improved bias decoupling was warranted. A drawing of the package used for these measurements is shown in Fig. 5-4. V-coax beads provide the interface between the external and internal environments. Internally, the coaxial pin is epoxied to a 50 Ω microstrip transmission line on a quartz substrate. This material was selected due to its excellent performance at high frequencies. The thickness of the quartz was selected to match that of the GaAs die to facilitate the wire bonding process. This package was designed and manufactured at Farran Technology. The associated assembly of the full amplifier was also carried out at the same location.

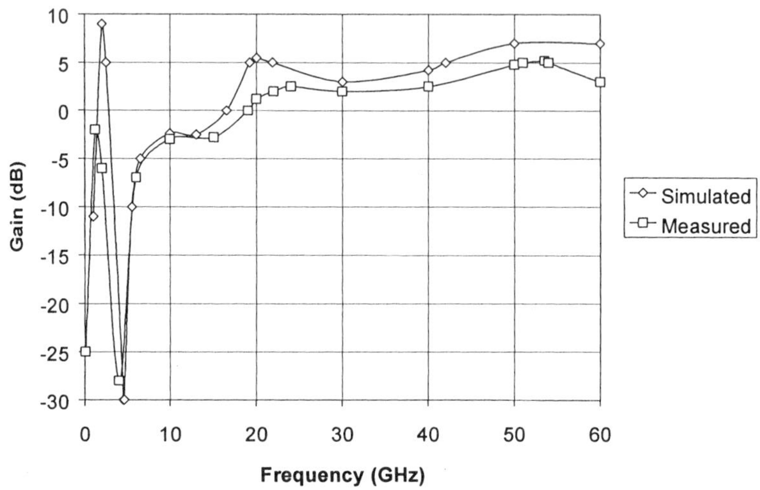

Figure 5-3. On-wafer measured and simulated gain vs frequency responses for single-stage 57 GHz LNA.

No oscillations were observed when the packaged amplifier was characterised. The optimum measured performance in the packaged environment was a gain in excess of 14 dB near 57 GHz with an associated noise figure of 12 dB. It was found that the gates needed to be slightly forward-biased to achieve this gain performance. A comparison of the simulated and measured gain for the two-stage amplifier is presented in Fig. 5-5. The predicted trace was simulated for more negative gate biases, consistent with both a lower DC current requirement and a lower noise figure. The predicted noise figure was less than 5 dB.

2.4 Discussion

It is clear that the measured performance falls short of the simulated, in particular in terms of noise figure (12 dB measured versus about 5 dB simulated). The measurement has a narrower-bandwidth gain response than that simulated. A possible contributor to these discrepancies is the imperfect extrapolation of the foundry device models (based on extractions from measurements to 40 GHz for s-parameters, and to 26 GHz for noise parameters) to the frequency range of interest. The more positive V_{gs} in the measurement relative to simulation (with an associated higher supply current) would be consistent with an increased noise figure – this increased current would also tend to increase the measured gain. Nevertheless, the measured gain response curve is sufficiently close to that simulated, in terms of the frequency range where the circuit functions best, to suggest that the design is reasonably sound and certainly shows promise both in terms of the circuit suitability for a 57 GHz radio link, and also in terms of the foundry process capability at V-band. To the authors' knowledge, this circuit represents the highest frequency amplifier developed on the H40 process.

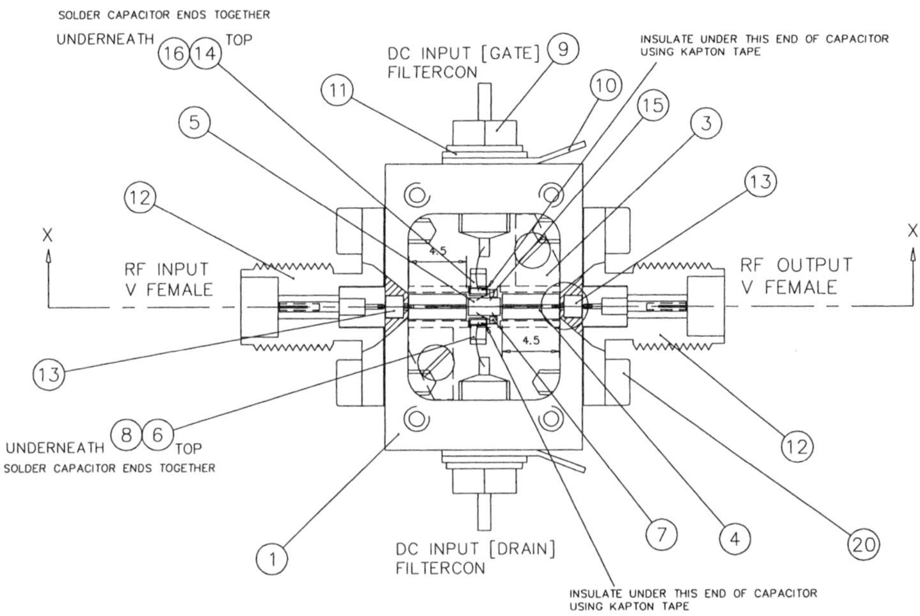

Figure 5-4. Outline drawing of assembly used for characterisation of the two-stage 57 GHz LNA.

However, two issues are clear. Firstly, there is risk associated with a circuit design so far outside the frequency range for which the models are supported by the foundry. Secondly, the response has quite a narrow bandwidth. This is mainly attributed to the design approach adopted which in essence involved placing 'blind' faith in the simulation tool output and involved very little in terms of intuitive understanding during the circuit optimisation process.

The remainder of this chapter addresses the latter issue. A sound method of LNA design is developed which is intuitive and facilitates the realisation of a broadband mm-wave circuit with excellent performance; moreover, the performance was achieved with a first pass design.

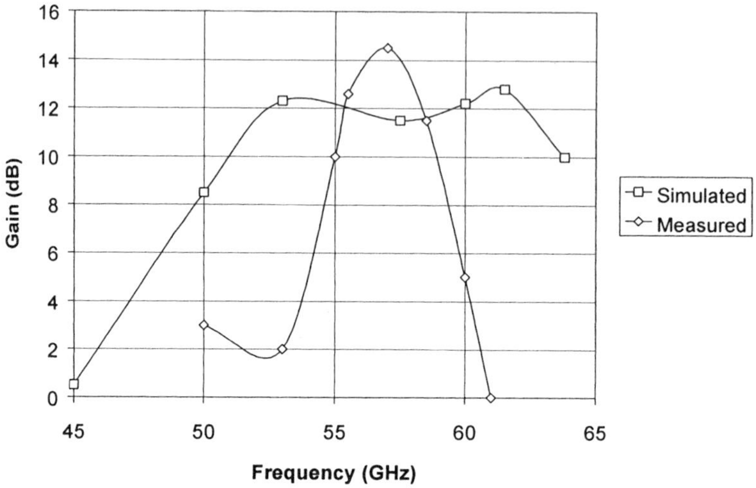

Figure 5-5. Measured and simulated gain vs frequency responses for two-stage 57 GHz LNA (measured data not corrected for front-end package loss).

3. NEW LNA DESIGN METHODOLOGY

This new methodology was developed in the context of a 40 GHz amplifier requirement. The target circuit is a standalone amplifier with input and output bondwire connections to 50 Ω terminations. The same methodology can also be applied to a transceiver chip requirement where the output load does not involve a bond-wire.

3.1 Considerations at mm-waves

Some of these considerations were taken into account in the 57.2 – 58.2 GHz LNA design earlier, but not in a systematic way. In this section, they are integrated as part of the new design methodology.

At high frequencies, the active device capability is constrained. In particular, gain tends to be at a premium, especially at reduced currents. As a consequence, the typical 10% of I_{dss} bias point for low noise performance is not appropriate due to a severe gain reduction in this bias regime. This is a key issue facing mm-wave LNA designers. There is no benefit to be gained in designing a LNA with very low noise figure if the associated gain is low because, when used in a receiver, the noise contributions of subsequent elements in the chain would be more significant due to the lower gain.

As a consequence, the gain/noise figure trade off needs to be studied in more detail, and the selection of the optimum bias point requires very careful consideration. The minimum *noise measure*[55], M_{min}, is a useful figure of merit for such a study. It is a single number that takes account of the minimum noise figure and associated gain of an active device. The minimum noise measure is derived from the theoretical noise figure of an infinite cascade of a given stage designed for minimum noise figure. The resulting expression is:

$$M_{\min} = F_{cascade} - 1 = \frac{F_{\min} - 1}{1 - 1/G_A}$$

where F_{min} is the minimum noise figure and G_A is the available gain of the individual stages. To a first-order, a reasonable assumption to make is that the lower the minimum noise measure, the better the potential performance of the device from a noise figure/gain compromise perspective. A summary of the predicted noise measure, at 40 GHz, as a function of current, for a 40 μm gate width H40 pHEMT device, is presented in Table 5-2. The simulations used the small signal foundry model for the pHEMT device.

Table 5-2. Noise Measure as a function of device current for a 40 μm pHEMT device

% I_{dss}	Noise Measure (40 GHz)
10	1.79
20	1.3
30	1.22
40	1.3
50	1.54

Similar trends are obtained for other device peripheries. On the basis of this analysis, it was concluded that a bias point of approximately 30% I_{dss} was optimal for 40 GHz low noise amplifier design. Further reductions in current do indeed yield improved minimum noise figure, but the corresponding gain degradation is excessive.

3.2 Device Stabilisation

Device stability is always a major concern in amplifier design, due to the potential for oscillation. Stabilisation techniques have been discussed at length elsewhere[9], and will not be covered here. However, mention will be made of the stabilisation *rationale* adopted in this work.

Essentially, there are two approaches to circuit stability. Firstly, one can bias the active device and design the embedding RF circuit to optimise the in band response to achieve the required performance specifications. On completing this effort, a stability analysis can be carried out and any potential for oscillation can be assessed and suitable fixes implemented on the previously designed circuit. The outcome is a circuit which may have its in-band simulated performance significantly degraded relative to the originally optimised solution. A number of iterative steps may now be necessary to correct the in-band response while at the same time ensuring stability.

An alternative view is to take the device and to modify it by combining it with suitable circuitry so that the resultant 'modified device' is unconditionally stable. Then, this modified device can be used for the RF circuit design, with the knowledge that the optimised circuit will also be unconditionally stable and will not require further stabilisation remedies.

This latter approach has been adopted for this work, particularly because it requires only one significant RF circuit optimisation effort, and such an effort at mm-wave frequencies can be considerable.

A schematic for the modified device is presented in Fig. 5-6. For the purposes of the mm-wave LNA design described in this work, a 2x60 μm device periphery was used. This device size provides a good trade-off between power handling capability and high frequency performance.

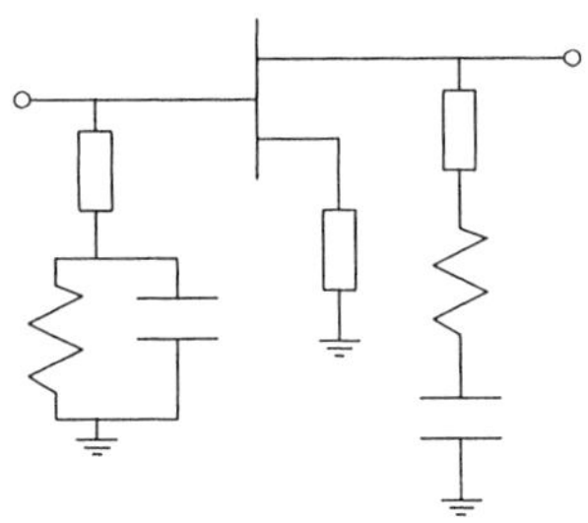

Figure 5-6. Schematic of stabilised device.

There are two stability networks to be seen here. At the gate, a parallel RC shunt arm to ground aids device stability at low frequencies. In the wanted band, the capacitor dominates and the resistor is decoupled from the circuit. This is important to minimise the influence of the gate resistor on the noise figure. The transmission line length to the RC combination is chosen to optimise the input match appropriately. This gate stability network is implemented as the gate bias circuit.

The second stability network is a series RC shunt arm on the drain. This network enhances the device stability at high frequencies (at low frequencies, the capacitor behaves like an open circuit, and the influence of the shunt arm is effectively eliminated) and has an influence on the in band frequency response of the circuit, essentially by introducing a lossy path to ground. As a consequence, this stability network must be connected at the drain rather than the gate side of the device in order to minimise its impact on the noise figure.

This modified device is unconditionally stable, and is used as the 'device' for subsequent amplifier designs in this work.

3.3 Packaging Considerations

To make a monolithic amplifier of interest for mm-wave applications, its potential for system integration must be considered. The conventional way of interfacing between MMICs and external components is by means of bond-wires. At higher frequencies, where the inductive reactance of such bond-wires can become an issue, some alternative packaging approaches are sometimes used. These approaches include ribbon bonding and flip-chip techniques. However, the use of wire bonding is preferred, where possible, due to its greater maturity as a process.

The inductance associated with a bond-wire is about 0.2 nH (typical for a 0.3 mm length of 25 μm gold wire). The inductive reactance associated with this inductance is 50.3 Ω at 40 GHz. In order to achieve good broadband performance from a circuit at this frequency, this reactance must be resonated out. Two options can be considered – series or shunt capacitive compensation.

A series capacitor of 79 fF would be required to resonate out the inductive reactance at 40 GHz. To realise such a small series capacitor would be extremely difficult, as parasitics would be significant and any deviation from the desired value would impact on the effectiveness of the compensation at 40 GHz.

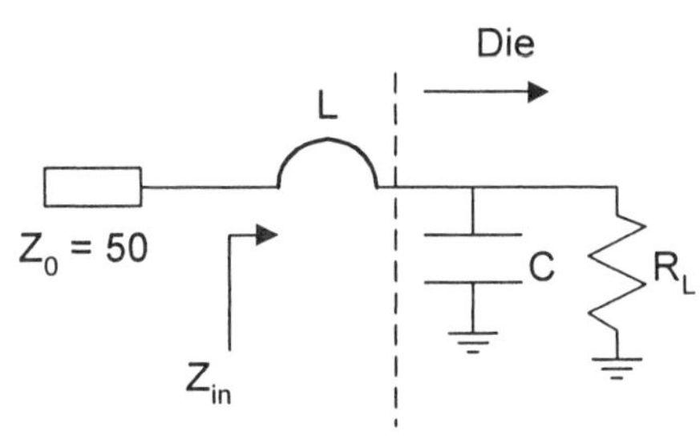

Figure 5-7. Bond-wire compensation using shunt capacitance on die.

An alternative approach is the use of a shunt capacitor on die. Such a capacitor can be implemented as a low impedance microstrip transmission line section. This situation is depicted in Fig. 5-7. Here, the bond-wire inductance is L, the compensating shunt capacitance on die is C, and the input impedance of the circuit beyond the capacitor is R_L.

Clearly, we can deduce that

$$Z_{in} = j2\pi f L + \frac{R_L (1/j2\pi f C)}{R_L + 1/j2\pi f C}$$

$$= j2\pi f L + \frac{R_L}{1 + j2\pi f C R_L}$$

$$= j2\pi f L + \frac{R_L (1 - j2\pi f C R_L)}{1 + (2\pi f C R_L)^2}$$

$$Z_{in} = \frac{R_L}{1 + (2\pi f C R_L)^2} + j\left[2\pi f L - \frac{R_L^2 2\pi f C}{1 + (2\pi f C R_L)^2}\right]$$

For f = 40 GHz, L = 0.2 nH and Z_{in} = 50 Ω, it follows that $R_L \approx 100$ Ω and $C \approx 40$ fF. Such a capacitance can be conveniently realised in the form of a wide low impedance transmission line section of suitable length. The simulated frequency response for a 100 Ω load matched to 50 Ω by the above arrangement is presented in Fig. 5-8. The net outcome of this analysis is that a good broadband match can be achieved to a MMIC with a bond-wire connection to the die included in the design. For this to be effective, the impedance looking into the amplifier circuit on die (beyond the capacitor) must be 100 Ω rather than the conventional 50 Ω.

To facilitate the on-wafer characterisation of an MMIC amplifier, ground-signal-ground pad arrangements can be incorporated at the input and output. The standard H40 probe-pad arrangements exhibit the very useful characteristic of being low impedance structures and it is found that they can fulfill the role of the shunt capacitance C in Fig. 5-7.

3.4 Noise/Match Tradeoff

Due to the high frequencies involved, the input impedance of a mm-wave FET is relatively low (at lower frequencies, it is a high impedance). This is because the input circuit of a FET consists primarily of the capacitance C_{gs} whose associated reactance becomes low at high frequencies.

Since the impedance seen looking into the signal source (i.e. looking from the load R_L back towards the input in Fig. 5-7) is about 100 Ω, and this impedance is significantly greater than the device input impedance, an effective way of presenting the device with a suitably low impedance match is to incorporate a quarter wavelength section between the 100 Ω point and the device input. This quarter-wavelength section transforms the 'large' 100 Ω impedance to a 'smaller' impedance at the device input. Given the constraints of microstrip and mm-wave GaAs technology, the minimum realisable characteristic impedance for this quarter-wavelength section is about 40 Ω. This results in a minimum resistive impedance, which can be presented to the device, of about 16 Ω, corresponding to a reflection coefficient $\rho = 0.515$ in a 50 Ω system.

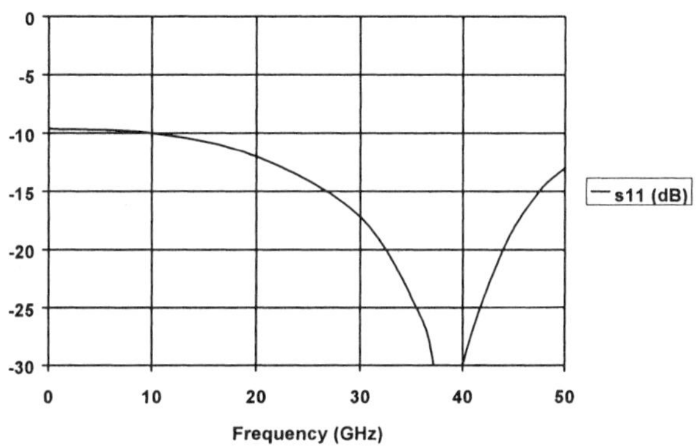

Figure 5-8. Frequency response of bond-wire compensation network.

Our goal is an LNA design with good performance at 40 GHz and with as broadband a characteristic as possible. The approach adopted was to nominally centre the design above 40 GHz – the motivation behind this is the certainty that broadband performance is always easier to realise subsequently at lower rather than higher frequencies. A comparison of the optimum conjugate match and Γ_{opt} for the modified device at 45 GHz is shown in Fig. 5-9. Also shown on this Smith chart are the 0.2 dB noise and the –0.3 dB gain circles. These intersect, and the $\rho = 0.515$ circle runs through this intersection region.

Hence, a suitable point, Γ_{comp}, on this $\rho = 0.515$ circle is an excellent compromise for the impedance presented to the input of the modified device, since it results in a noise figure which is no more than 0.2 dB degraded from its minimum possible value, and a gain that is less than 0.3 dB below its optimum value. This corresponds to an input return loss of at least 11.7 dB. Thus, with this approach, a very good tradeoff of noise figure versus match ought to be possible. The reactive component of the required impedance can be introduced by means of a small phase rotation from the real axis impedance presented by the quarter wave transformer.

A similar noise/match tradeoff at 50 GHz for the same modified device is presented in Fig. 5-10. A similar comment applies here in terms of the ρ = 0.515 circle being consistent with both good noise figure and match. In this case, the impedances have almost no reactive component, thereby indicating that practically no phase rotation is required between the input quarter wave transformer and the input of the modified device.

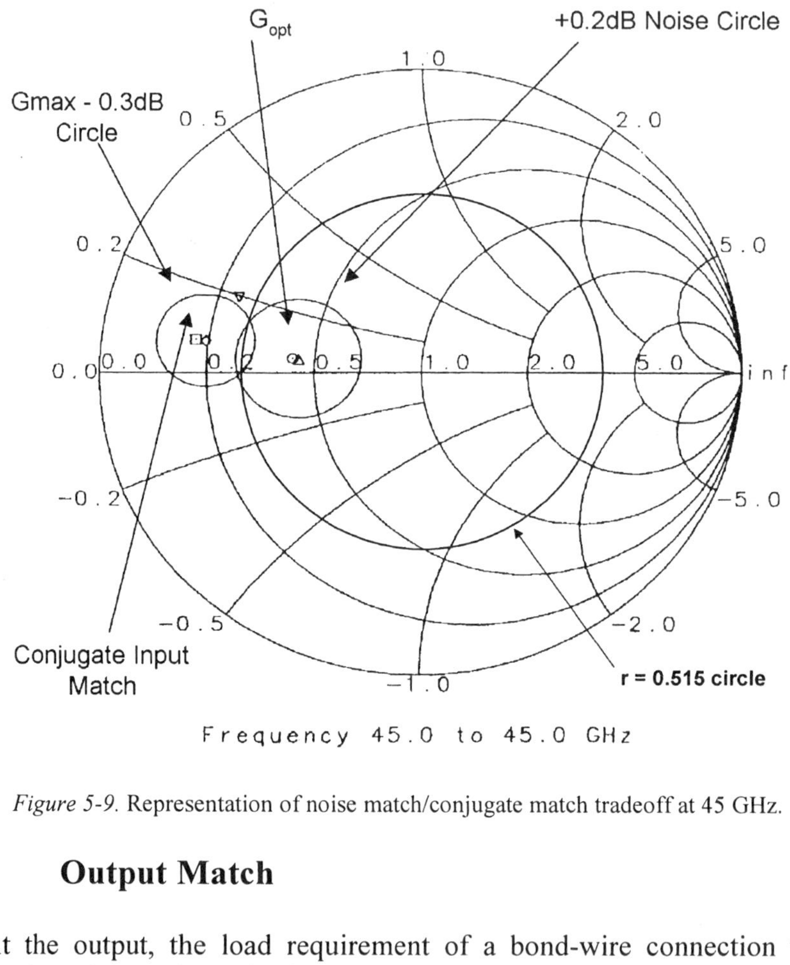

Figure 5-9. Representation of noise match/conjugate match tradeoff at 45 GHz.

3.5 Output Match

At the output, the load requirement of a bond-wire connection to an external 50 Ω impedance is similar to the input interface described earlier. A similar use of a low impedance section of line (i.e. shunt capacitance) is appropriate. As a consequence, the effective load impedance on die prior to the low impedance transmission line section is 100 Ω.

When the modified device is presented with the input circuit outlined in the previous section, it is found that, at 45 GHz, the reflection coefficient seen looking back into the drain of the device is Γ_2 = 0.52∠-157 °. This impedance point is remarkably close to the ρ = 0.515 circle. The net result is that a small phase rotation is sufficient to rotate this impedance around to the real axis. From this real impedance, a similar quarter wave section as used in the input can be used to transform the impedance to 100 Ω.

It should be noted here that integrating the amplifier in a transceiver design (without an output bond-wire requirement) would simply require that a different impedance output quarter-wavelength section be used. The purpose of this alternative impedance would be to change the output impedance of the amplifier from 100 Ω to 50 Ω.

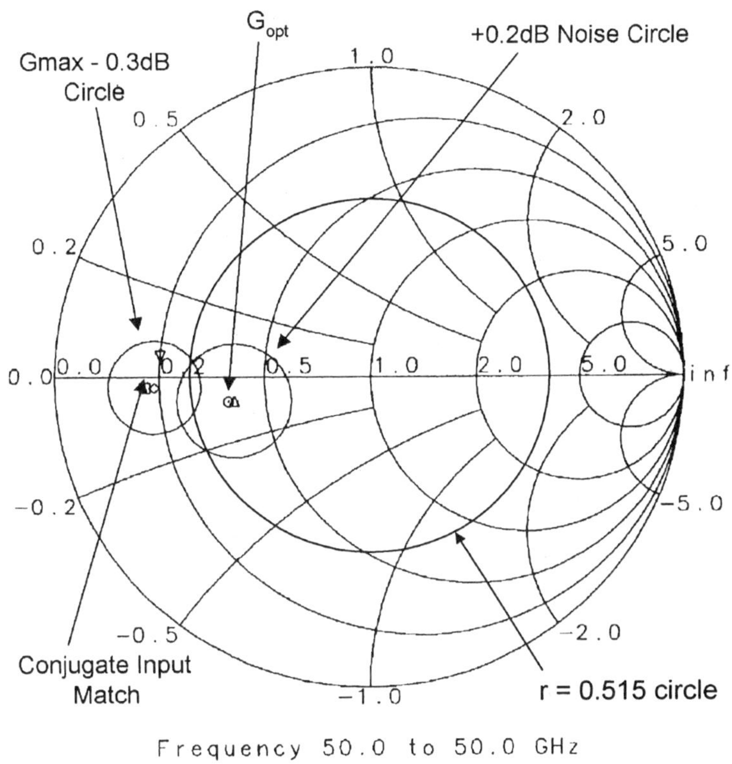

Figure 5-10. Representation of noise match/conjugate match tradeoff at 50 GHz.

3.6 Inter-stage Match

Interestingly, it is also clear that instead of rotating to the real axis point, it is also possible to rotate, again using a short length of line, from Γ_2 to the Γ_{comp} point. In this way, it becomes apparent that the modified device can be cascaded with itself easily, with the inter-stage match consisting only of a short length of line. This is the case for designs developed for both 45 GHz and 50 GHz. This is a very interesting feature of the design. The ease of cascading means that the modified device can be regarded as a standard building block which can be cascaded as often as necessary to satisfy a given gain requirement.

3.7 Amplifier Bandwidth Enhancement

In addition to the broad-bandwidth bond-wire compensation networks, the simulated bandwidth of the circuit was further broadened by the inclusion of suitable short circuit stubs at the input and output. This was an extension of the double stub method which, in its basic form, uses a parallel combination of a short circuit and an open circuit stub, each of length $\lambda/8$ at the centre frequency. The admittances associated with these stubs cancel at the band centre, but can introduce a useful reactive load over frequency to improve the achievable match away from the centre of the band. In this work, it was found that the short circuit stub on its own proved very effective in broadening the simulated bandwidth of an amplifier. Figs. 5-11 and 5-12 show the predicted performance of a three stage 30 – 50 GHz amplifier before and after the inclusion of the broad-banding stub.

3.8 Four stage 30 – 50 GHz LNA

The predicted small signal performance for a four-stage amplifier, developed by cascading stages as explained above, is shown in Fig. 5-13. The simulated compression characteristic at 45 GHz is presented in Fig. 5-14. From this characteristic, a predicted output P1dB of approximately +11 dBm is inferred.

The circuit layout for the four-stage LNA is shown in Fig. 5-15. Here, it is immediately apparent that the individual amplifier stage building blocks are compact, with the main die area being dominated by the quarter wave matches at the input and output. As a consequence, the addition of new stages has a relatively small effect on the overall die area, and the *gain per square-mm* figure of merit is high. The die size is 2.73 mm x 1.53 mm. The die is laid out with a view to on-wafer probing, a key requirement for production testing in volume applications where the die would typically be marketed and sold in the bare die form, rather than necessarily being packaged. To further enhance the suitability for on-wafer probing, the DC input pads have also been laid out with adjacent grounds with a standard footprint pattern to facilitate DC probe cards with built-in bias decoupling.

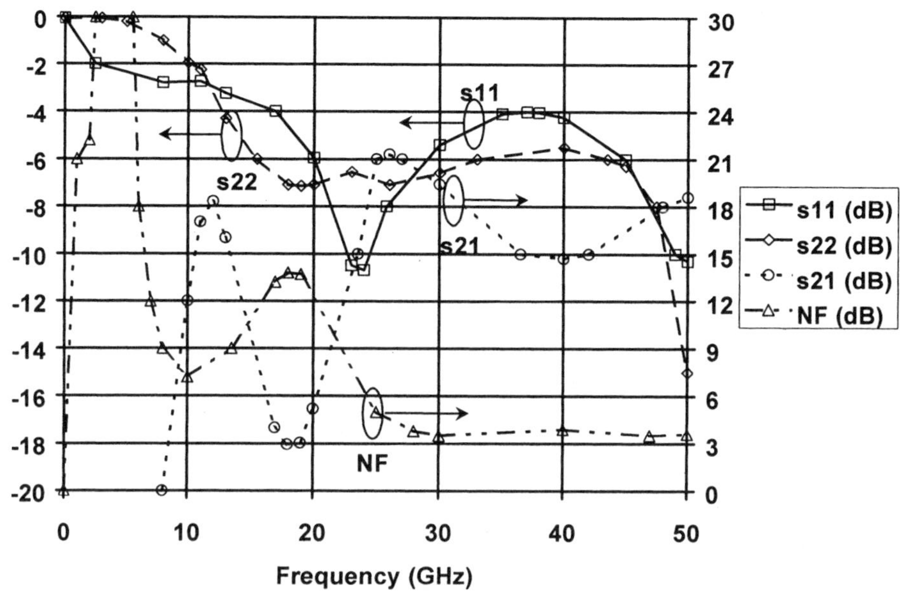

Figure 5-11. Simulated response for three-stage LNA without broad-banding stub.

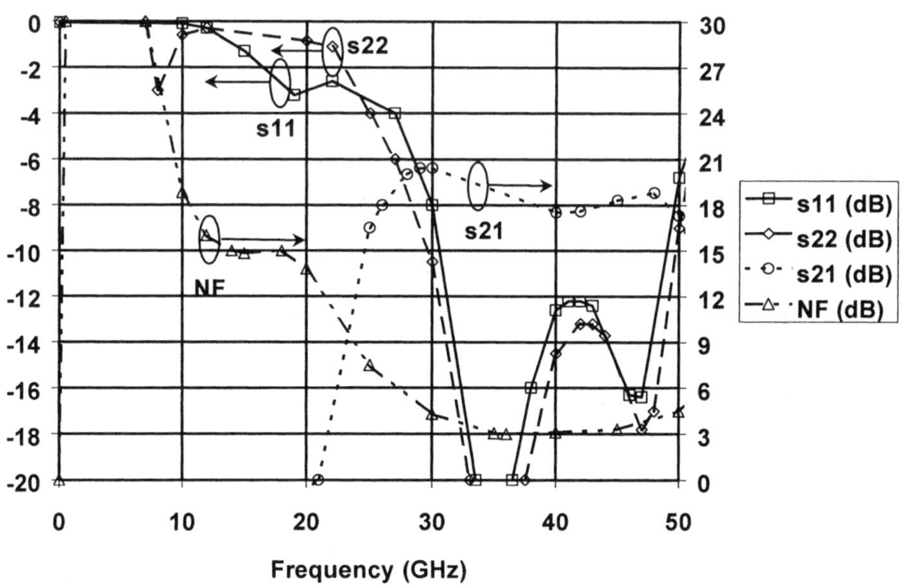

Figure 5-12. Simulated response for three-stage LNA with broad-banding stub.

3.9 Measurements

The four-stage LNA was assembled in a customised test package. The concept consisted of the die being epoxied to the base of a metal block. The RF interface to the chip consisted of short wire bonds to 50 Ω microstrip transmission lines on Rogers Duroid 5880[56]. The duroid is also epoxied to the base of the block. The external RF connections are provided by coaxial connectors in the side of the block. The DC pads on the MMIC are wire bonded to adjacent chip decoupling capacitors. These are in turn wire bonded to a flexible bias control PCB that facilitates bias tuning for the various bias connections. A photograph of the assembled test package is shown in Fig. 5-16.

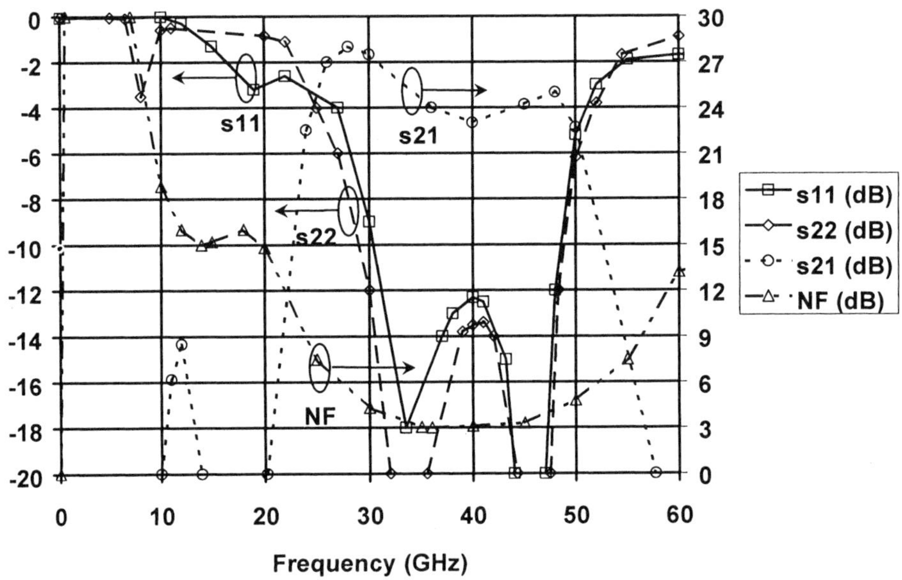

Figure 5-13. Simulated response for four-stage 30 – 50 GHz LNA.

The packaged LNA was then tested using an Anritsu-Wiltron 54177A scalar network analyser. The bias voltages were tuned for optimum performance in a broadband sense. The resulting bias conditions were: V_{DD} = 2.22V, V_{g1} = -0.19V, V_{g2} = -0.27V, V_{g3} = -0.09V, V_{g4} = -0.15V, I_{DD} = 114mA. The measured s-parameters are presented in Fig. 5-17. The measured response shown includes the loss due to the package, and this is estimated at 2 dB at 40 GHz. Taking account of the package loss, the inferred MMIC measured gain exceeds 20 dB for the frequency range 26 –

44 GHz. The corresponding frequency range for the simulated gain is approximately 23 – 51 GHz. A comparison of the simulated and measured (corrected for loss) amplifier gain over the frequency range 20 – 50 GHz is shown in Fig. 5-18.

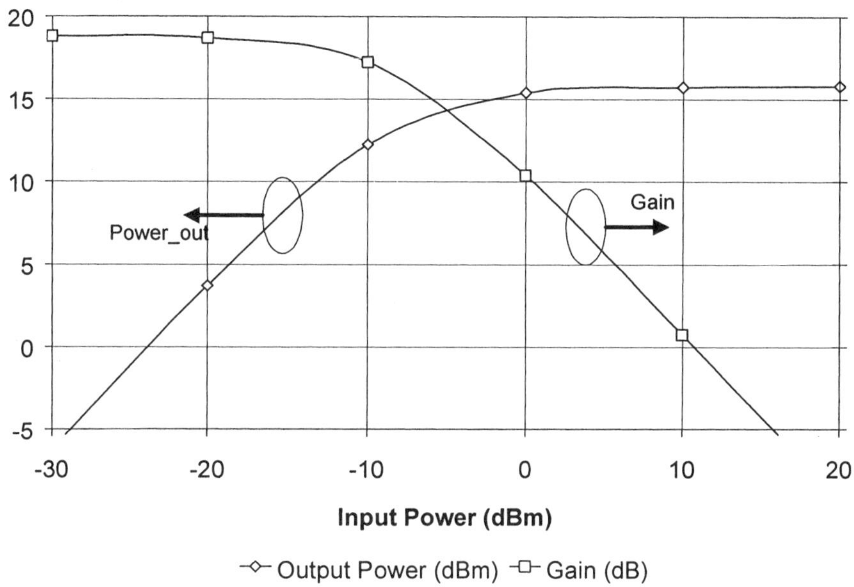

Figure 5-14. Simulated compression characteristic for four-stage 30 – 50 GHz LNA.

The correlation between the measured and simulated responses is highly encouraging, in particular at lower frequencies. There is a peak in simulated gain near 48 GHz that is not replicated in measurement. It is possible that this is due to packaging loss becoming a constraint at high frequencies, but the cause could also be due to the use of device models extracted to 40 GHz and extrapolated for simulations at significantly higher frequencies. Nevertheless, the measured performance validates the design approach, and it is clear that the MMIC provides significant gain over a broad range of frequencies.

In terms of noise figure, the LNA was characterised in the same package using a HP-8971B noise figure test set. The mm-wave tests were made possible by the additional use of a Farran Technology Ltd. BDC-28S down converter module, providing noise figure measurement capability from 26.5 – 40 GHz. The measured LNA noise figure, de-embedded for front-end losses, is also shown in Figs. 5-17 and 5-18. A value of 4.5 dB is obtained between 35 and 40 GHz. The predicted noise figure over this frequency range is 3.1 dB. A number of factors could be contributing to this discrepancy. The most likely candidate is invalid noise models for the active devices – the H40 noise extraction was based on measurements to 26 GHz, and extrapolation to 40 GHz is likely to be suspect. Nevertheless, the achieved performance of 4.5 dB results in a high performance amplifier that is suitable for use over a broad frequency range. It is also evident in Fig. 5-18 that the measured noise figure response follows the same trend over frequency as that suggested by simulation, and therefore further upholds the design approach.

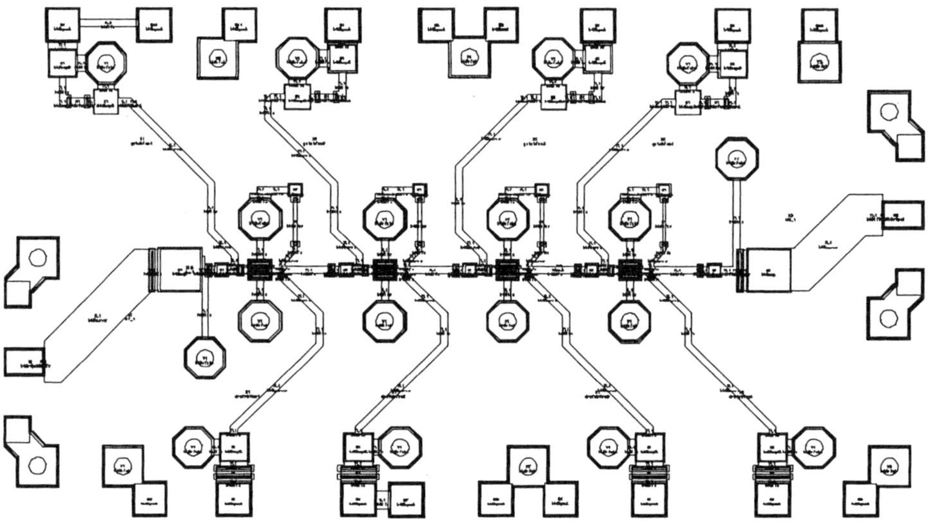

Figure 5-15. Four-stage 30 – 50 GHz LNA layout (2.73 mm x 1.53 mm).

The compression characteristic of the amplifier was also evaluated in the package, and the output P1dB was measured to be +10 dBm. Allowing for package loss, the estimated P1dB for the MMIC is +11 dBm. This agrees remarkably well with the simulated value as inferred from Fig. 5-14. All stages in the LNA use a 2x60 μm pHEMT device and thus the output power equates to approximately 105 mW/mm.

4. DISCUSSION

The 57 GHz LNA design presented in Section 2 of this chapter was developed using conventional techniques. Of course, great care was taken in the design to ensure the maximum likelihood of yielding a functional mm-wave circuit. From this perspective, the design was successful and provided excellent gain in the frequency range of interest. However, its noise figure was significantly higher than expected, and the DC current draw was also somewhat excessive. It is believed that this discrepancy is mainly due to the use of device models with limited accuracy in the frequency range of interest.

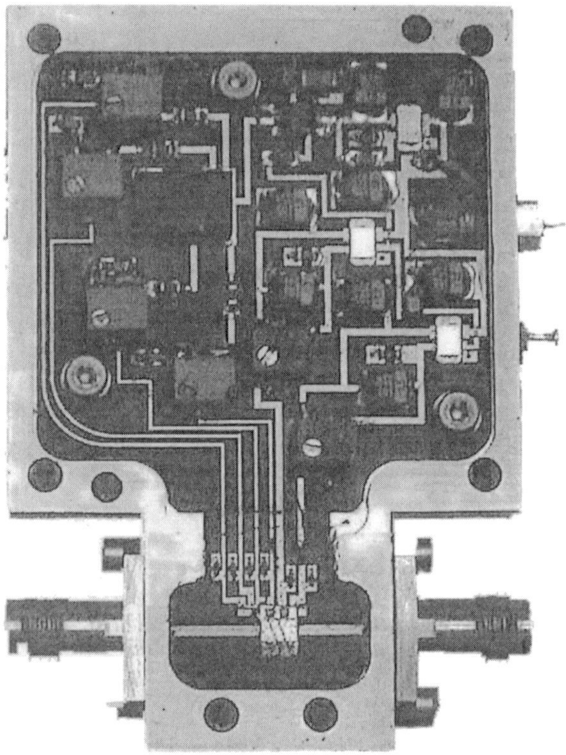

Figure 5-16. Photograph of assembled MMIC 30 – 50 GHz LNA in customised package.

The novel LNA design methodology presented in Section 3 of this chapter has resulted in a fabricated mm-wave amplifier with excellent performance over a broad frequency range. The design approach consists of firstly stabilising the active device such that its subsequent use in the amplifier design does not warrant further stabilisation efforts. The device stabilisation was carried out in such a way as to minimise the likely impact on the final amplifier noise figure. This was achieved by ensuring minimal in-band resistive loading at the input of the active device. Having stabilised the device, the next step involved presenting the input with a compromise impedance to get the optimal noise figure versus return loss tradeoff. A simple, powerful and compact realisation of the compromise impedance has been demonstrated. The neutralisation of the input and output bond-wires through the use of low impedance probe pad structures has been outlined, and this makes possible a very broadband circuit response consistent with the mature wire bond interface technology.

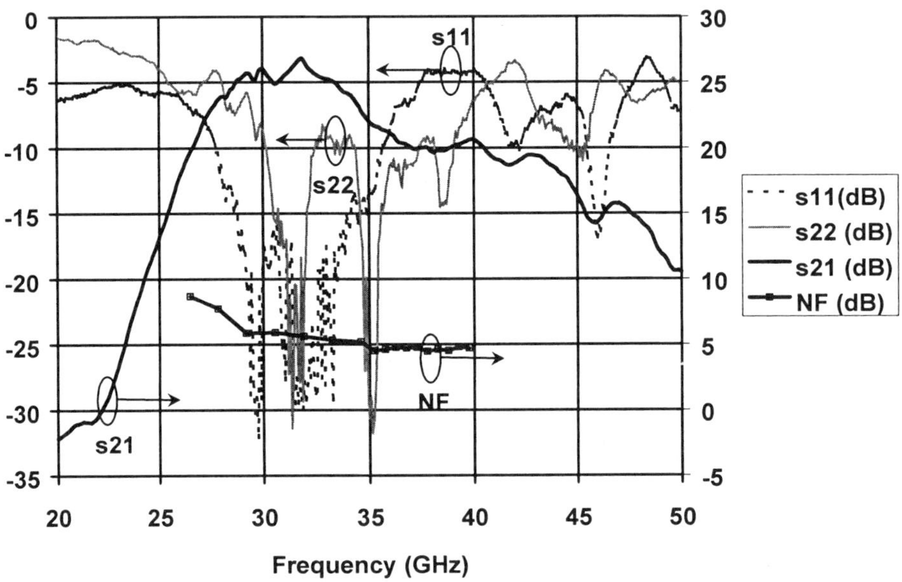

Figure 5-17. Measured response for four-stage LNA (s-parameters in package; NF de-embedded for front-end package loss).

The resulting amplifier MMIC has been characterised in a customised metal package design, and the results validate the design approach. A true broad-band gain response has been achieved, with very useful noise figure and compression characteristics. The MMIC is certainly suited to integration in mm-wave subsystems operating in the frequency range 25 – 45 GHz. The overall agreement between the simulated and measured amplifier performance is impressive, with a good correlation being realised for the gain, the noise figure and P1dB.

Some discrepancies do exist between the measured and simulated responses, in particular above 45 GHz. This is not entirely surprising given that the H40 model extractions are based on foundry characterisations to 40 GHz; extrapolation to higher frequencies is likely to be suspect.

Nevertheless, the measured performance certainly demonstrates that the objective of achieving a broad-band amplifier response with a low noise figure has been achieved. The design approach presented is particularly suited to mm-wave amplifiers, due to the requirement for only one significant mm-wave response optimisation effort after the initial device stabilisation step.

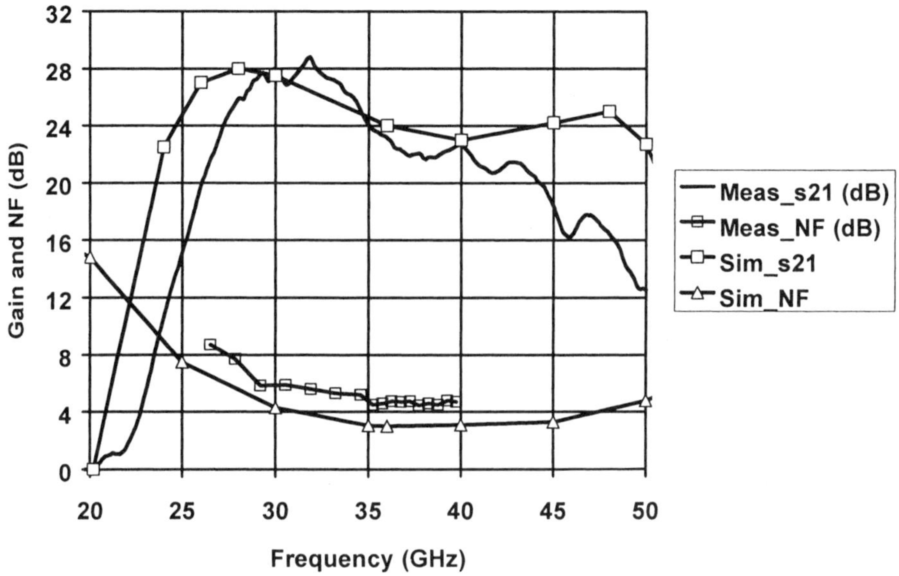

Figure 5-18. Comparison of measured and simulated performance for four-stage LNA.

5. CONCLUSION

A novel LNA design methodology has been introduced in this chapter. Application of this approach results in a very compact amplifier stage that facilitates convenient cascading with an excellent gain per unit area figure of merit.

The initial device stabilisation step requires only one subsequent circuit response design effort, and this is of particular benefit at mm-waves due to the design and simulation complexity – the alternative stabilisation approach which involves stabilising the circuit *after* the amplifier response has been optimised inevitably results in an iterative design process with more than one optimisation step.

The new design methodology has been validated by the measurements carried out on a 30 – 50 GHz LNA which functions very well, with good gain and noise figure performance being realised over a broad bandwidth; this LNA was designed using the new approach. The benefits associated with the use of the new design technique are best illustrated by a comparison of the performance of the 30 – 50 GHz circuit with the corresponding performance of the 57.2 GHz amplifier in Section 2 of this chapter, which was not designed using the new approach. The design methodology comprises a useful new aid to the mm-wave circuit designer. The new methodology also lends itself to both standalone amplifier design, as in the case of the four-stage LNA described in this chapter, and also to integrated transceiver amplifiers by appropriate choice of the impedance of the output quarter-wavelength section.

Chapter 6

MONOLITHIC MIXERS

1. INTRODUCTION

The frequency-converting mixer is a key element of any mm-wave transceiver subsystem[6]. The mixer plays the essential role of converting between the low intermediate frequency (IF) and the mm-wave radio frequency (RF). Historically, high frequency mixers have been based on Schottky diodes, and have usually been implemented in waveguide technology[57]. More recently, with the emergence of FETs with good high frequency performance, significant progress has been reported in the field of FET based mixers[58]. The emergence of viable MMIC mm-wave mixers is a key driver for the commercialisation of the mm-wave frequency range through the development of low cost transceiver products suitable for high volume production.

In this chapter, some of the key considerations in FET mixer design will be studied. A number of FET mixer topologies will be discussed, and their performance analysed intuitively. The optimum biasing conditions will be assessed for these topologies. The advantages of a balanced approach are outlined. Design details on a novel balanced FET mixer developed during the course of this work for a 57 GHz transceiver application (see Chapter 4), and fabricated on the GMMT H40 process will be presented. The measured performance will be discussed, and compared with the corresponding simulations.

In addition, a Schottky diode based balanced mixer, also intended for the same 57 GHz requirement, has been designed and fabricated on the H40 process. The novel feature here is that the diode layout is based on the FET cell layout with a short circuit applied between the source and drain terminals. Details of this diode mixer circuit will be discussed, and measured parameters will be presented, again in the context of expected performance. The limitations associated with a Schottky diode being implemented on HEMT material, and also the limitations due to the use of a pinchoff-extracted FET model for the diode, will be considered, especially in the context of the measured performance of the mixer.

The relative merits of the FET and diode mixer approaches will be discussed, and the selection of the balanced diode mixer for the 57 GHz transceiver application will be justified in terms of its reasonable performance as both an up converter and a down converter. Moreover, the requirement of an off-chip IF hybrid for the balanced FET mixer compromises its suitability for use in an integrated transceiver.

Finally, state of the art developments will be discussed, and the chapter will close with a discussion of likely future trends.

2. MICROWAVE MIXERS – SOME CONSIDERATIONS

A mixer is a non-linear three-port component. When such a component is driven by multi-tones, intermodulation products are generated and, by means of suitable circuit design and filtering, the required mixing product can be extracted. A diode is very suitable for use in such a component due to its highly non-linear I-V characteristic. A Schottky diode is particularly suited to high frequency applications. A FET is less non-linear but, due to its active nature, it offers the potential for some conversion gain, particularly at lower frequencies. This has meant that a FET based mixer can replace a conventional diode mixer and amplifier in certain applications. However, to work well, the biasing of a FET mixer is critical, and the particular bias point depends on the mixer topology adopted.

A mixer is typically represented as shown in Fig. 6-1. The three ports are usually termed the RF (radio frequency), the LO (local oscillator) and the IF (intermediate frequency) ports. For a down converter, the inputs are the RF and LO, and the output is a lower frequency (IF) which is typically the difference between the RF and LO frequencies. In the case of an up converter, the inputs are the IF and LO, and the output is the RF which is typically the sum of the IF and LO frequencies.

Usually, the LO signal is at a higher power level than the others. This signal 'pumps' the mixer and causes it to be non-linear, thereby introducing frequency mixing effects when other tones are present. By careful selection of the operating bias point of a mixer, the LO input requirement can be minimised. For a down converter, the RF signal is usually at a much lower

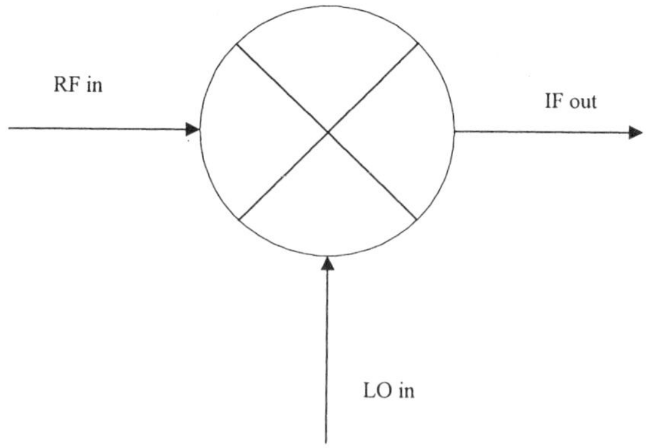

Figure 6-1. Conventional mixer representation.

power level, this signal often being derived directly from an antenna perhaps via a low noise amplifier. The output IF signal is also normally at a relatively low level. The ratio of the IF and RF power levels is termed the conversion loss, usually expressed in dB. Typically, the IF is at a lower level and the conversion loss is expressed as a positive number of dB. In the case of high specification FET based mixers, the output IF level can actually exceed that of the RF and the mixer is said to have conversion gain.

For an up converter, the LO level is similar to that of a down converter. In this case, the output RF is generally lower in power level than the applied IF, and the ratio of the RF to the IF is termed the conversion loss.

The mixing products in a mixer are generated as a consequence of some non-linear characteristic. In a typical diode mixer, for example, the non-linear diode I-V characteristic plays a key role. When two or more tones are applied to a device with a non-linear characteristic, the individual tones are multiplied together, thereby producing sum and difference frequencies. The detailed theory behind the generation of these mixing products has been presented by others, and Maas[6] is recommended to the interested reader.

Given a non-linear device, the next concern is the operating bias point. The primary purpose of the bias point it to optimise the non-linearity of the device to ensure maximum mixing and optimum conversion efficiency. For example, in the case of a Schottky diode device, the LO power requirement can be reduced considerably by operating the diode with a slight forward bias. This trade off between design complexity (and external bias source requirement) and performance is typical of many high frequency developments. In the case of a FET mixer, both the drain and gate bias voltages must be considered. Many different FET mixer schemes have been reported[59] and some of these have been studied in this work.

The noise figure of a mixer sets a limit on its sensitivity to low RF levels. Below a certain RF power level, the intrinsic noise generated within the mixer itself drowns the RF and renders the mixer incapable of reception. The noise performance is optimised through careful design. This includes optimum biasing, and to the extent possible, the elimination of thermal noise sources (such as resistors).

Another performance parameter of a mixer is its linearity in terms of intermodulation. This involves mixing of two or more RF tones, and is undesirable as it may fall within the wanted band and may not be filterable. For example, the third order intermodulation product associated with two closely spaced RF tones would typically lie in band – this is a key specification requirement of many mixers under the guise of IP3. As the RF level increases, such intermodulation becomes more of a concern. Eventually, when the RF levels are high enough, the intermodulation output will exceed the desired IF. Mixer design often incorporates multiple individual mixing devices in a balanced configuration with a view to reducing (through cancellation) some of the intermodulation products and also with a view to improving the isolation between the mixer ports. The intermodulation characteristic sets the maximum RF level for which the mixer provides reliable reception.

The mixer noise figure and intermodulation distortion characteristics combined determine its dynamic range. This corresponds to the range of RF levels for which the mixer provides useful performance.

Having selected the optimum bias point for the mixing device, the other design inputs involve the input and output embedding or matching networks. The input network(s) plays two key roles. Firstly, the input signal becomes matched into the mixing element. For the case of a down converting gate or transconductance FET mixer, where both the LO and RF are combined and applied to the gate, the matching network aims to provide a good compromise match for both signals. This network also provides a reactive load to the IF such that any IF present at the input side of the mixer is reflected back into the mixer and is not terminated. Similarly, the output network is designed to extract the maximum IF power and, at the same time, to reactively terminate the high frequency signals and reflect them back into the mixing element. Essentially, the overall mixing efficiency is improved by reactively terminating the unwanted signals at the mixer ports. Similar considerations apply for an up converter.

3. MONOLITHIC MIXER ARCHITECTURES

3.1 Single-Ended Diode Mixers

The I-V characteristic of a Schottky diode is the primary source of non-linearity in a diode mixer (another source being the Q-V characteristic that is particularly exploited in varactor type frequency multipliers). The relationship between the current and voltage can be represented by a power series:

$$i(t) = a_0 + a_1 v(t) + a_2 v^2(t) + a_3 v^3(t) + \ldots$$

Assuming that the LO and RF signals (in the case of a down converter) are combined and then applied across the diode, $v(t) = v_{LO}\cos(\omega_{LO}t + \phi_{LO}) + v_{RF}\cos(\omega_{RF}t + \phi_{RF})$ can be substituted in the I-V power series. This immediately yields a whole series of mixing products. Thus the frequencies of these mixing products can be defined for all positive and negative integers m, n:

$$\omega_{m,n} = m\omega_{LO} + n\omega_{RF}$$

$\omega_{m,n}$ is also denoted by (m, n). For the typical situation where $v_{LO} >> v_{RF}$, only the terms for $n = 1$ are significant. When the RF level increases, higher order terms begin to increase rapidly in magnitude.

In the case of a conventional down converter, the difference frequency corresponding to either $(1, -1)$ or $(-1, 1)$ is the desired IF, depending on whether the LO is above or below the RF.

The LO power requirement of a diode mixer can be minimised by suitably biasing the device. This shifts the operating point up the I-V curve to make the instantaneous current more sensitive to the changes in instantaneous voltage. An interesting approach adopted here, in particular at mm-wave frequencies, is to bias the mixer diode with a constant current source. The voltage applied in order to maintain this constant current is then monitored and the down converted IF is extracted from this voltage signal by low pass filtering.

Due to the predictable non-linear characteristic associated with a diode, the frequency conversion properties of a diode mixer have been widely characterised and understood[6].

MMIC diode mixers have been reported. In the work by Viaud[60] for example, the design of a number of key components of a mm-wave transceiver is described. In particular, up and down converting diode mixers are outlined. Both circuits have been developed on a 1 μm United Monolithic Semiconductors (UMS)[48] Schottky diode process, dedicated specifically to mm-wave low conversion loss mixer circuits. This point is worth emphasising. In order to achieve good conversion performance from a diode mixer, UMS has found that the best approach has involved the development of a process for this specific requirement, rather than attempting to use its well established pHEMT process for this purpose.

3.2 Sub-Harmonically Pumped Diode Mixers

In the transceiver topology studied in the course of this work, the difficulty in providing a high frequency LO source directly is addressed in Chapter 7 by means of a frequency multiplier between the LO input and the mixing element. An alternative technique which is commonly applied, in particular at even higher frequencies, and often using waveguide technology, is the sub-harmonically pumped mixer. In this mixer, the external LO is applied at a sub-harmonic of the effective LO, and the internal mixer circuit design is arranged such that the frequency of the LO signal is effectively multiplied. Most sub-harmonic mixers are designed for a harmonic number of two. Such a mixer is often implemented using a pair of anti-parallel Schottky diodes mounted across a waveguide.

Sub-harmonic mixers have also been reported in MMIC technology[61]. This sub-harmonic mixer[61] is integrated with a V-band LNA and a LO buffer amplifier on a 0.18 μm pHEMT process. The f_{max} of this particular process is 200 GHz, and its I_{max} is 550 mA/mm. The mixer is implemented using a pair of 12.5 μm Schottky diodes which have the same (anode) recess structure as the gate of a pHEMT device. The design consists of a $\lambda_{RF}/2$ short circuit stub on the side of the diodes opposite where the RF is applied. This stub presents a short to the RF and at the same time presents a high impedance (close to an open circuit) to the LO input. A $\lambda_{LO}/4$ open circuit stub is located at the end opposite the LO injection. This stub presents a short to the LO, and an approximate open circuit to the RF. In fact, the two stubs are approximately the same length, with the difference being dependent on the IF. The short circuit stub is implemented by means of a capacitor to ground, and the IF is extracted from the non-grounded end of the capacitor. This is shown conceptually in Fig. 6-2.

For good subharmonic mixer performance, the diode pair must be well matched[6]. This is of course a reasonable expectation of a good MMIC process, and hence it is likely that such designs will feature more and more in future developments as high frequency MMIC processes mature. However, the sub-harmonic mixer was not considered for the developments in this work, primarily because the intended application lent itself more to a

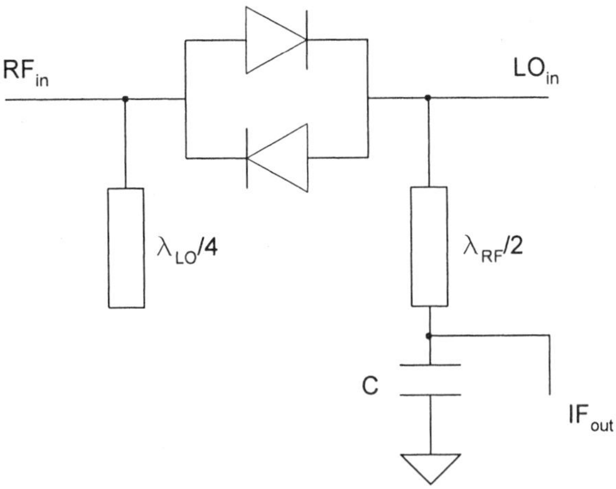

Figure 6-2. Conceptual sub-harmonic mixer realization.

frequency multiplier approach on the LO and, moreover, the foundry did not offer a quality diode model for its process. It is the opinion of the authors, that sub-harmonic mixers will emerge as key elements of future mm-wave MMIC transceiver architectures. This fact is further supported by the work of Ellis[62]. That work discusses the potential for higher order (odd harmonic) mixers. The mixer circuit in this case consists of a single shunt diode. Ellis demonstrates through simulation that by appropriate design of the planar circuit, the conversion loss can remain relatively constant across a wide range of pump frequencies. He also demonstrates the viability of his approach by means of a scale model at 25 GHz, with a fifth sub-harmonic of the LO being applied. Ellis also makes the crucial point that conventional high frequency diode mixers based on beam lead Schottky diodes are performance constrained due to the diode packaging parasitics. Significantly improved performance is to be expected through the use of MMIC technology. In fact, Ellis presents a simulation of a W-band 9th sub-harmonic mixer with a predicted conversion loss in the 15 – 20 dB range. Such performance would compare favourably with that typically expected from the waveguide based subharmonic mixers of today.

3.3 FET Mixers

FET based mixers have become increasingly popular in recent years. In this context, the FET mixer is used to describe mixers based not alone on the conventional MESFET, but also on other field effect transistor devices such as the pHEMT. In this work, for example, the development of a balanced down converting pHEMT mixer is described. The terms FET mixer and pHEMT mixer will be used interchangeably in the remainder of this work. The FET offers some distinct advantages over the diode from a mixer standpoint. Firstly, the FET is a three-terminal device and, in some configurations, the three mixer ports can be associated with the three FET connections. In a diode mixer, this is clearly not possible. This can give the FET mixer some advantage in terms of port to port isolation. A FET mixer also offers the potential for conversion gain. This is specifically the case for an active FET mixer, and such conversion gain has often been reported for relatively low frequency designs. Passive FET mixers have also been developed. These rely on some aspect of the variable resistance (between drain and source) characteristic of a FET under varying gate bias control, and such circuits inevitably involve conversion loss rather than gain.

A key distinction between a FET and a diode is that the non-linear regime of a FET is very much bias dependent, whereas a diode characteristic is non–linear throughout. As a consequence, the biasing of a FET mixer requires much more careful consideration than a diode mixer in order to obtain useful performance. Depending on the topology adopted, and the port connections to the FET terminals, the device characteristic is most non-linear near pinchoff in some cases, in the saturation region in others, and the linear region in other scenarios. As a consequence, a major element of FET mixer design is the bias point selection.

4. FET MIXERS – PRIOR ART IN THE CONTEXT OF THEORY AND PRACTICAL REALISATION FROM THE LITERATURE

A vast amount of research has been published on the topic of mixer theory. Both diode and FET mixers have been analysed in considerable detail[6]. There is little to be gained by reproducing this theory here, but a few key points will be made in advance of the original development work being presented in this chapter.

In terms of understanding the fundamental performance of FET mixers, the *transconductance* mixer is a suitable vehicle as it is probably the most common FET mixer topology. In this arrangement, both the LO and RF signals are applied to the gate, with the IF being extracted from the drain. The LO, being applied directly to the gate, directly modulates the FET's g_m. The time varying g_m can be represented by a Fourier series[58]

$$g_m(t) = a_0 + \sum_{1}^{\infty} a_n \cos(n\omega_{LO} t),$$

where the Fourier coefficients have their usual meanings. Given that g_m is defined as

$$g_m = \frac{\partial i_{ds}}{\partial v_{gs}},$$

for small RF levels, the AC component of the drain current can then be reduced to

$$i_d(t) = g_m(t) v_{RF}(t)$$

It is clear from an examination of these equations that the IF difference frequency component is maximised when the a_l coefficient is maximum – see Section 5.2.1.1 of this chapter. An interesting analysis of this situation is presented in Schefer et al[58]. For a HEMT device, the relationship between a_l and bias/LO level is illustrated graphically. This figure is reproduced here as Fig. 6-3.

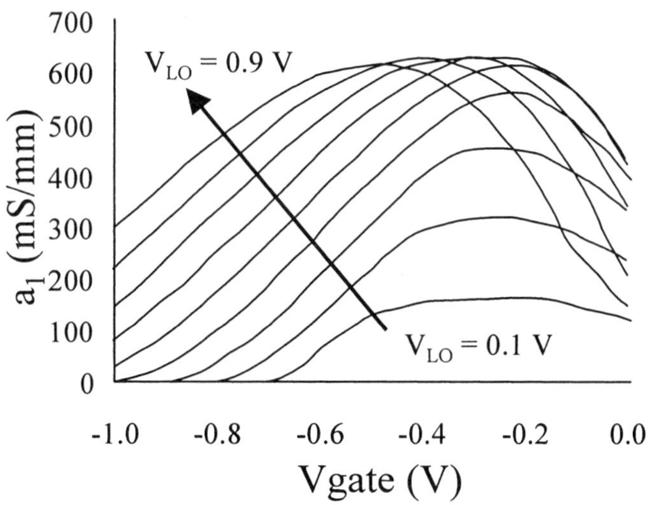

Figure 6-3. a_1 versus bias/LO drive (source Schefer et al[58], © 2004 IEEE).

The general conclusion drawn[58] is that the optimum gate bias voltage, in terms of IF conversion performance, is in the linear region of the transfer characteristic, between pinchoff and where the transconductance is maximum. The precise bias point for peak a_1 is a function of the LO level, with increasing LO powers being associated with a_1 peaks at more negative gate biases.

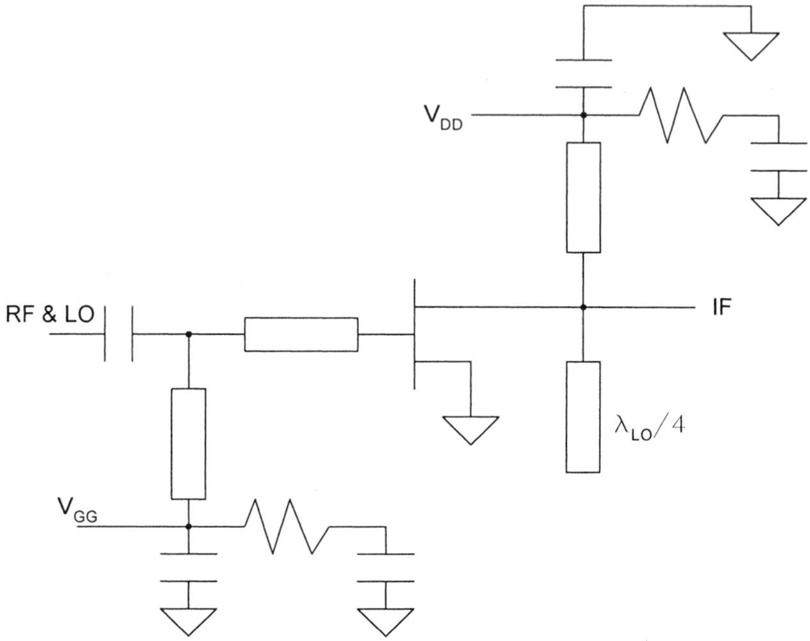

Figure 6-4. Transconductance mixer schematic.

A similar analysis to that above can be employed in principle for all FET mixer topologies. Such analyses will be presented in an intuitive manner in the following section.

A V-band HEMT transconductance mixer design is also described by Schefer et al[58]. A simplified schematic for this circuit is presented in Fig. 6-4. The LO short circuit at the device drain is provided by a quarter wavelength open circuit stub. This suppresses the LO at this point, and ensures that the device operates in the saturation region over the complete LO cycle by maintaining $V_{DS}(t)$ close to V_{DD}. This in turn maximises the gain potential of the circuit. The resistors in the bias networks are present for stability purposes. The decoupling capacitors directly to ground are relatively low in value and are effective at the RF/LO frequencies. The capacitors in series with the resistors are much larger in value and are present to ensure that the low frequencies are decoupled to ground via the resistors. This helps to eliminate the potential for bias oscillations. This schematic comprises a very useful template for a practical transconducance mixer.

5. ORIGINAL MIXER STUDY, DESIGN AND EVALUATION

In the development of MMIC mixers for the 57 GHz transceiver requirement, a number of topologies were initially studied. These were studied firstly as very basic single-ended circuits and were compared primarily on the basis of the predicted conversion performance. The device model used was a scalable pHEMT model extracted by the foundry specifically for pinchoff type biasing. Having identified the preferred topology, the single-ended design was then refined by the inclusion of all the appropriate foundry elements to make the design realisable. Other aspects of a mm-wave mixer design, critical to a design's success, will be introduced. Balanced mixer options will be considered, and the advantages of such implementations will be discussed. This entire design evolution, and the specific design and simulated performance of the balanced FET mixer topology selected for fabrication will be presented in the following sections.

In addition to the FET mixer development, a diode mixer has also been designed. The approach adopted will be described, and the balanced design generated through this work will be outlined. The FET and diode mixer designs will be compared and contrasted.

The mixers studied during the course of this work were simulated using the HP-Eesof (Agilent) Libra CAE tool[27]. The topologies studied were investigated in terms of their suitability for the 57 GHz radio link application. For the purposes of this requirement, both up and down converting mixers are of interest.

In very simple terms, the structure of a commonly used FET mixer design can be represented as shown in Fig. 6-5. This is shown conceptually as a transconductance style mixer (see Section 5.2.4 of this chapter), with both the RF and LO applied to the gate. However, the principles apply equally well to other FET mixer topologies.

A topology which can in principle present the desired embedding impedances was shown schematically in Fig. 6-4, which for convenience is repeated as Fig. 6-6, with additional detail in terms of component identification.

For good mixing operation, the embedding impedances Z_1 and Z_2 should meet the following conditions:

On the input, the shorted stub (TL_2) is used to present a low impedance at the IF. This requires a relatively large decoupling capacitor C_1, or alternatively a DC short could be used with a self-bias arrangement on the source of the FET. This shorted stub is used for RF/LO tuning. Where low frequency oscillations may be of concern, C_1 can be reduced in value (in order to remain effective at mm-wave frequencies) and the additional resistor and capacitor (R_1 and C_2) can be introduced for stabilisation. C_1 also provides gate bias decoupling.

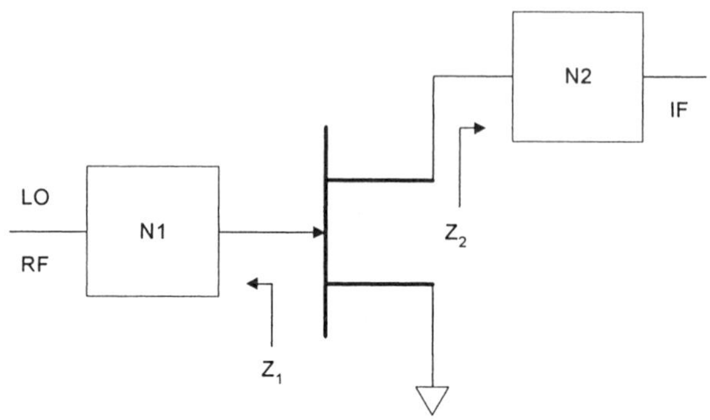

Figure 6-5. Basic FET mixer representation.

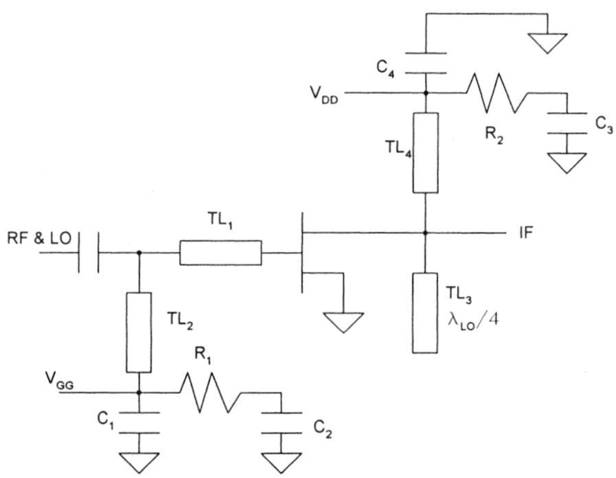

Figure 6-6. Transconductance mixer schematic (as per Fig. 6-4).

On the output side, the quarter-wavelength TL_3 open circuit stub presents a low impedance to the RF and LO signals which leak to the drain terminal. This stub is practically extraneous at low intermediate frequencies. The other stub, TL_4, is used in conjunction with C_4 to enhance the IF match. C_4 also provides drain bias decoupling. The R_2 and C_3 combination can again be utilised to enhance stability.

Thus far, the mixer embedding impedances have been discussed with no clear indication of how the FET ought to be biased. For optimum performance of a FET mixer, the bias point must be chosen appropriately for the particular topology adopted. This will be discussed in some detail, from a fundamental perspective, in Section 5.2 of this chapter.

5.1 Simulation Tool Settings Specific to Mixer Design

5.1.1 Harmonic and Mixing Order Settings

The mixer is a non-linear circuit, and its simulation using many CAE tools involves what is termed the *Harmonic Balance* approach[6]. This analysis method takes the form of a mix of frequency domain (for the linear constituents of the circuit) and time domain (for the non-linear constituents) analyses. The transformation between the two domains involves Fourier analysis. Essentially, an iterative analysis is applied until such time as the solution converges.

For the harmonic balance engine to work, the number of harmonics of the various input frequencies considered in the analysis must be defined. This variable, NH, is set by the designer. For highly non-linear circuits, or high power levels, this number should be set relatively large, whereas for weakly non-linear circuits, and/or low power levels, the harmonic number can be set low. In fact, for the limiting case where the power levels are very low, the analysis reduces to a linear analysis and the harmonic value can be set to unity. Of course, in the case of a mixer, such a reduction is absurd as no mixing action is then possible in the simulation!

In the analysis phase, the selection of an appropriate harmonic factor setting is a matter of trial and error. Start with a relatively low value (e.g. NH = 3) and check the simulated response. Then increase the value and continue in this way until the simulated results have converged. Larger values are likely to yield a more accurate prediction, but at the expense of increased simulation time and computer memory usage, and as a consequence it is prudent to accept the lowest possible setting which is consistent with achieving the right result.

5.1.2 Mixer Mode On/Off

The Mixer Mode can be set to either *On* or *Off*. It applies only to mixer circuit analysis. If this mode is *On*, the simulation tool assumes that for the multi-tones applied to the circuit, only one is of sufficient power level to warrant inclusion of its harmonics. All other applied tones are treated as small signals and only their fundamental frequencies and first order mixing products are taken into account in the simulation. On the other hand, if the mode is set to *Off*, the simulation tool treats all the applied tones as large signals and all their harmonic products, up to the maximum number set by the NH factor above, are included in the analysis.

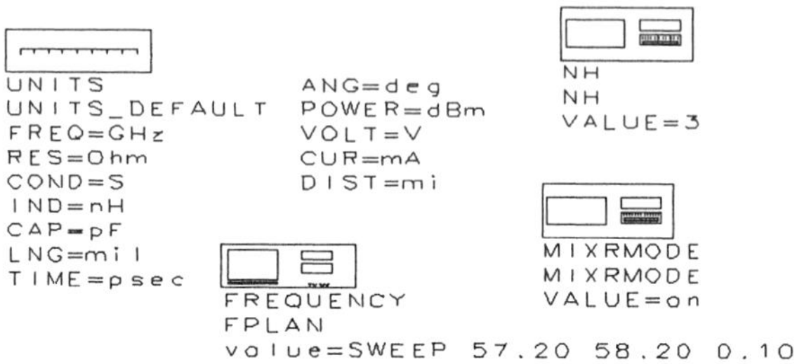

Figure 6-7. Typical elements of Agilent CAE[27] test bench for harmonic balance simulation.

The mixer mode being *On* is advantageous from the standpoint of simulation speed as fewer frequency terms are taken into account in calculating the solution. On the other hand, it would reasonably be expected that the mode being *Off* would yield a more accurate solution. However, there is a growing consensus in the design community that the Mixer Mode *On* may in fact yield the more accurate conversion loss response. Clearly, it does not take all the possible intermodulation effects into account, but there is anecdotal evidence that it does give results which are more consistent with measured results[63]. It is not clear why this is the case, but it is probably related to some form of truncation error when an excessive number of mixing products are computed.

Every effort has been made in this work to develop circuits that yield reasonable simulated responses using both simulation modes.

Some of the above items are contained in an example Harmonic Test Bench, shown in Fig. 6-7.

5.2 Single-Ended FET Mixer Topologies – Intuitive Study and 57 GHz Down Converter Simulations

In the course of this work, a number of single-ended FET mixer topologies have been studied. Depending on the port connections, different aspects of the non-linear characteristics of a FET are exploited to generate the mixing products. The evolution of these designs will now be discussed in detail. Firstly, the FET mixer topologies are introduced, and these are considered intuitively from a down converting perspective. This intuitive analysis is the original work of the authors, and constitutes a simple but effective aid to the understanding of FET mixer operation. For up converting operation, a similar fundamental thinking applies. The circular markers in the I-V characteristics indicate the approximate optimum bias point for each design configuration, and the bold lines indicate the instantaneous bias trajectory over the LO cycle.

Some of these approaches have also been examined in more detail in the form of simulation studies, with the 57 GHz transceiver as the application requirement, using the H40 foundry process as the underlying technology. For each of the topologies considered, the appropriate simulation study is presented directly after the corresponding intuitive analysis.

5.2.1 TOPOLOGY #1 LO to drain, RF to gate

5.2.1.1 Generic Intuitive Study

In this configuration, the large signal LO drives the drain terminal, and the small signal RF is applied to the gate. The key output of interest is the IF component of the FET current, $i_{DS}(t)$. With the RF applied to the gate, the FET parameter of particular importance is g_m. This is apparent when the definition of g_m is considered:

$$g_m = \frac{\partial i_{DS}}{\partial v_{GS}}$$

With the RF applied to the gate, $v_{RF}(t)$ becomes associated with $v_{GS}(t)$ and the IF component of $i_{DS}(t)$ is derived from the mixing of $v_{GS}(t)$ and the LO component of $g_m(t)$. This is clearly the case from the following:

$$v_{GS}(t) = V_{RF}\cos(\omega_{RF}t + \phi_{RF}) + V_{DC}$$

$$g_m(t) = g_{m0} + \sum_{n=1}^{\infty} g_{mn}\cos(n\omega_{LO}t + \phi_{n)}$$

where $g_m(t)$ has been represented by its Fourier Series. For small variations in V_{GS},

$$\delta\, i_{DS} = g_m\, \delta\, v_{GS}$$

$$\delta\ \left(i_{DS}(t)\right) = \left(g_{m0} + \sum_{n=1}^{\infty} g_{mn}\cos(n\omega_{LO}t + \phi_{n)}\right) \bullet \delta\ \left(V_{RF}\cos(\omega_{RF}t + \phi_{RF}) + V_{DC}\right)$$

The difference frequency, IF, component of $i_{DS}(t)$ is derived from the mixing of the RF component of $v_{GS}(t)$ with the LO component of $g_m(t)$ according to

$$V_{RF}\cos(\omega_{RF}t+\phi_{RF})\bullet g_{m1}\cos(\omega_{LO}t+\phi_1)\approx I_{DS,IF}\cos(\omega_{IF}t+\phi_{IF})$$

As a consequence, the conversion from the RF to the IF is optimum when g_{m1}, the LO component of $g_m(t)$, is maximised.

With the LO applied across the drain – source, the key question then becomes where to bias the FET to ensure that g_{m1} is maximised, or specifically to ensure that the change in g_m with respect to V_{DS} is maximised.

Consider the sketch in Fig. 6-8 that shows the trends in both g_m and its derivative versus V_{DS}. As indicated in Fig. 6-8, the solid lines correspond to a low value of V_{GS}, while the dashed lines correspond to a higher V_{GS}. This shows clearly that in order to optimise the IF generation for this mixer configuration, the best bias point is in the linear region (low V_{DS}). g_m itself increases with increasing V_{DS}, buts its rate of increase lessens in the saturation region. V_{GS} needs to be significantly above pinchoff to enable current to flow and thus facilitate significant values for g_m.

The FET is biased near the knee of the I-V curve as shown in Fig. 6-9. The LO pump shifts the instantaneous bias point along an almost constant V_{gs} curve (with the small deviations from the constant trace being due to the instantaneous RF voltage).

TOPOLOGY #1 provides a convenient separation of the LO and RF inputs. The LO – RF isolation is enhanced by the reverse isolation of the FET device. Good RF – LO isolation may be difficult to achieve, and this could be an issue in some applications where frequency pulling of the LO source could result.

5.2.1.2 Practical Simulation Study

This approach would be problematic from a number of perspectives. Firstly, it would require diplexing of the LO and IF at the drain. Secondly, achieving good RF - LO isolation would be difficult, particularly in a single-ended design. For a low IF, the RF termination at the drain is likely to be non-optimum (i.e. non reactive due to the LO and RF frequencies being close) and achievement of good mixing performance would be constrained. As a consequence, this approach was not considered any further at this stage.

5.2.2 TOPOLOGY #2 LO to source, RF to gate

5.2.2.1 Generic Intuitive Study

Following the same line of thought as above for TOPOLOGY #1, where the RF is applied to the gate, the key FET characteristic is g_m, and the conversion performance is optimum when the rate of change of g_m with respect to the LO is maximised.

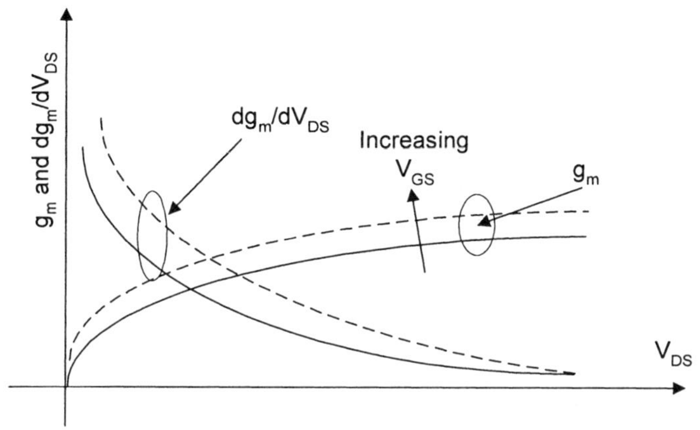

Figure 6-8. Representation of g_m and its derivative versus V_{DS} for FET.

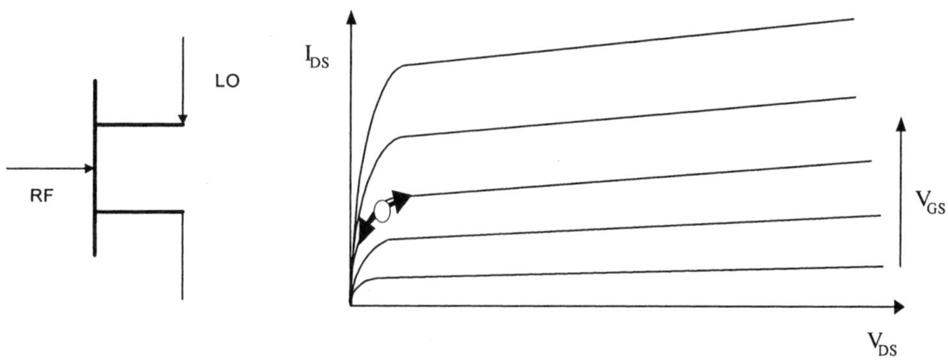

Figure 6-9. I-V Trajectory over LO cycle for TOPOLOGY #1 FET mixer.

However, with the LO applied to the source, it is clear that the LO modulates both V_{GS} and V_{DS}. Thus, the IF generation is impacted by the rate of change of g_m with respect to both V_{GS} and V_{DS}.

It has already been outlined that g_m changes most rapidly with V_{DS} in the linear region. It is also clear that g_m changes most rapidly with V_{GS} in the saturation region. Thus, for this configuration, the optimum bias point is in fact a compromise between two conflicting requirements.

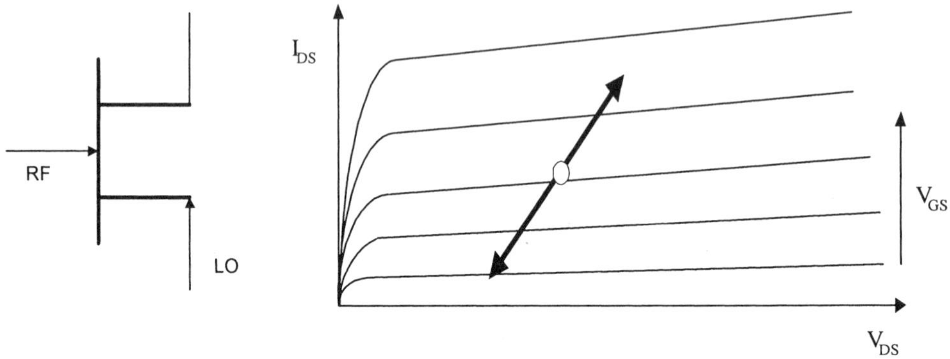

Figure 6-10. I-V Trajectory over LO cycle for TOPOLOGY #2 FET mixer.

The bias trajectory is shown in Fig. 6-10, with the compromise bias point indicated by the circle. For the H40 pHEMT device, the V_{GS} associated with this bias point is negative.

TOPOLOGY #2 will manifest good LO – RF isolation due to the reverse isolation characteristic of the FET. However, in this instance, the injection of the LO at the source conflicts with providing a good high frequency ground at that terminal. Moreover, due to the more complex nature of the LO modulation impact on the FET characteristics, the accurate simulation of its performance is likely to be a concern.

5.2.2.2 Practical Simulation Study

A crude first level schematic, used to assess the potential of this configuration, is shown in Fig. 6-11. The stub and large value capacitor C_2 to ground on the input equates to a short circuited stub for the RF and LO, and aid RF input matching. At the same time, the stub's effective electrical length at the IF is small and the capacitor effectively presents the desired low impedance at the IF at the device's input. If a self bias arrangement is possible, the gate can be tied to DC ground either by dispensing with the C_2 and short circuiting the stub directly to ground, or alternatively via resistor R_1. In order that the required negative V_{GS} be realised, the shorted gate would require a positive bias at the source, and this can be conveniently provided by means of a self-bias scheme. Of course, the self-bias network is relatively immune to bias tuning after fabrication, and a more flexible design in which the input stub is capacitively connected to ground with a large value capacitor and in which a non-zero gate bias can be injected is desirable for circuit evaluation and optimisation.

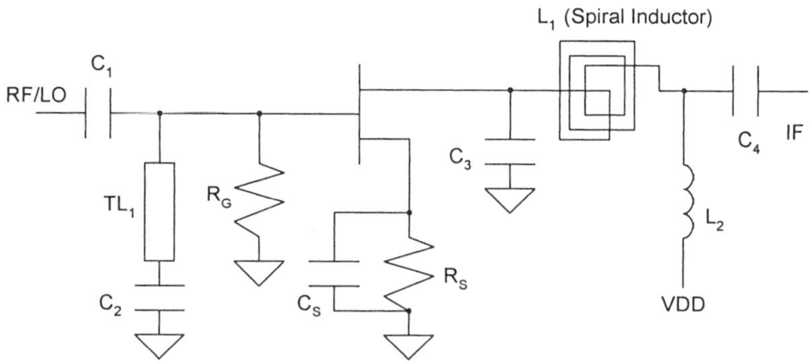

Figure 6-11. Initial TOPOLOGY #2 mixer schematic for simulation.

At the output (drain), the low pass filter prevents the RF and LO from breaking through to the IF port. The leading element of the low pass filter, shunt capacitor C_4, presents a low impedance at the drain to the high frequency signals. The inductor and capacitor (L_2 and C_5) adjacent to the IF port may not need to have such large values, and can possibly be incorporated into the IF matching structure. The implementation of series inductor L_1 in the low pass filter is straightforward at lower frequencies, but at mm-wave frequencies this becomes more of a problem due to the limited self-resonant frequencies of the typical spiral inductors supported by MMIC fabrication processes – above its self-resonant frequency, an inductor exhibits primarily capacitive characteristics. For certain, it would be imprudent to use the H40[3] standard element spirals where significant mm-wave signals exist as the frequency range of operation exceeds their self-resonant frequencies and their inductive properties are very suspect. In fact, it only makes sense to utilise such spirals in parts of the circuit which have been decoupled from the mm-wave signals by other means. Possible decoupling approaches will be discussed later.

Nevertheless, the simulation of the above 'idealised' design (including a spiral inductor representation of L_1) is interesting as it provides a quick estimate of the best case performance which this topology could possibly deliver before non-idealities and parasitics are factored in. Under these conditions, the simulated performance is summarised in Fig. 6-12.

Given that the typical conversion loss performance realised using a high performance mm-wave Schottky diode based waveguide mixer would be of the order of 6dB, this simulated FET mixer performance is very encouraging.

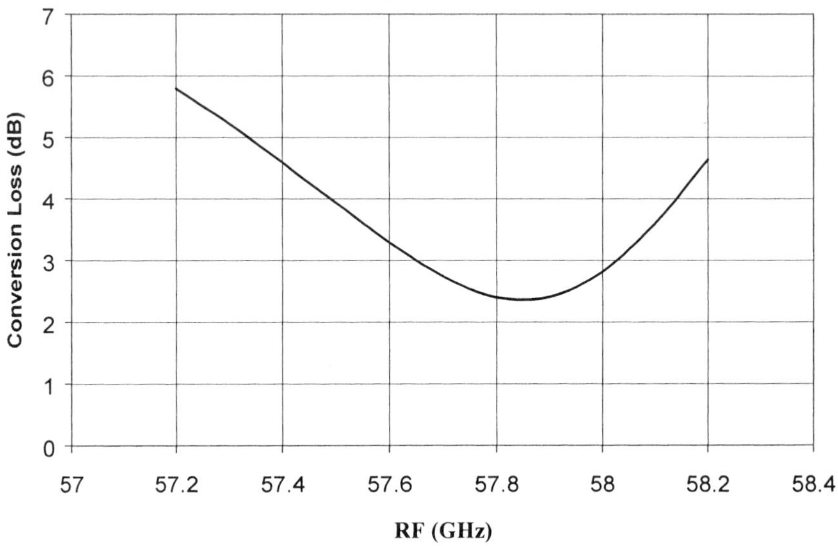

Figure 6-12. Initial TOPOLOGY #2 FET mixer simulated conversion loss for LO of +10 dBm at 56.2 GHz.

When the LO level is backed off from +10 dBm to 0 dBm for the same circuit, the predicted conversion loss degrades to 12.26 dB at 57.2 GHz and 11.18 dB at 58.2 GHz, a typical degradation of about 7 dB. This follows from the fact that the circuit was optimised for +10 dBm of LO drive, and deviations of the LO power either above or below the design level can degrade performance. This is unlike a diode mixer where, typically, performance will saturate and not degrade as the LO power is increased beyond a certain point.

5.2.3 TOPOLOGY #3 LO to gate, RF to source

5.2.3.1 Generic Intuitive Study

In this configuration, the large signal LO drives the gate terminal, and the small signal RF is applied to the source. Once again, the key output of interest is the IF component of the FET current, $i_{DS}(t)$. With the RF applied to the source, the FET parameter of particular importance is g_{DS}. This is apparent when the definition of g_{DS} and the Fourier series representation of this parameter are considered:

$$g_{DS} = \frac{\partial i_{DS}}{\partial v_{DS}}$$

$$g_{DS}(t) = g_{DS0} + \sum_{n=1}^{\infty} g_{DSn} \cos(n\omega_{LO} t + \phi_n)$$

With the RF applied to the source, $v_{RF}(t)$ becomes associated with $v_{DS}(t)$ and the IF component of $i_{DS}(t)$ is derived from the mixing of $v_{DS}(t)$ and g_{DS1}, the LO component of $g_{DS}(t)$. As a consequence, the conversion from the RF to the IF is optimum when g_{DS1} is maximised. With the LO applied across the gate – source, the issue becomes where to bias the FET to ensure that g_{DS1} is maximised, or specifically to ensure that the change in g_{DS} with respect to V_{GS} is maximised.

Consider the sketch in Fig. 6-13, which shows the trends in g_{DS} versus V_{GS}.

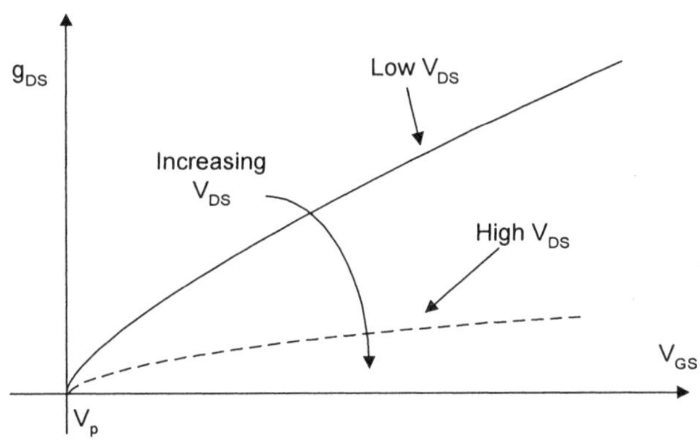

Figure 6-13. Representation of g_{DS} versus V_{GS} for a FET.

It is clear that in order to optimise the IF generation for this mixer configuration, the best bias point is again in the linear region (low V_{DS}), with a low value of V_{GS} - this will be referred to as TOPOLOGY #3a. This maximises the derivative of g_{DS} with respect to V_{GS}.

It is also apparent that significant variation in g_{DS} with respect to V_{GS} can be expected at higher V_{DS}, provided that V_{GS} is maintained near pinchoff – this will be denoted TOPOLOGY #3b. However, it is clear from the sketch that the rate of change in g_{DS} with respect to V_{GS} is greater for a bias point in the linear region.

5.2.3.2 TOPOLOGY 3a

In TOPOLOGY #3a, the FET is biased at $V_{DS} = 0$ and the LO drive shifts the instantaneous bias as shown in Fig. 6-14. As this is not an active bias point, such a FET mixer is intrinsically incapable of providing conversion gain. In this configuration, the FET behaves like a zero current switch, with the effective resistance presented between the drain and source switching between high and low values over the LO cycle (such a mixer is often referred to as a resistive mixer). It has a reduced LO power requirement when compared with a diode mixer. It also offers the potential for reduced LO leakage to the RF port. Variants of this approach are widely applied in RF balanced mixers where a differential RF is applied to both the source and drain terminals.

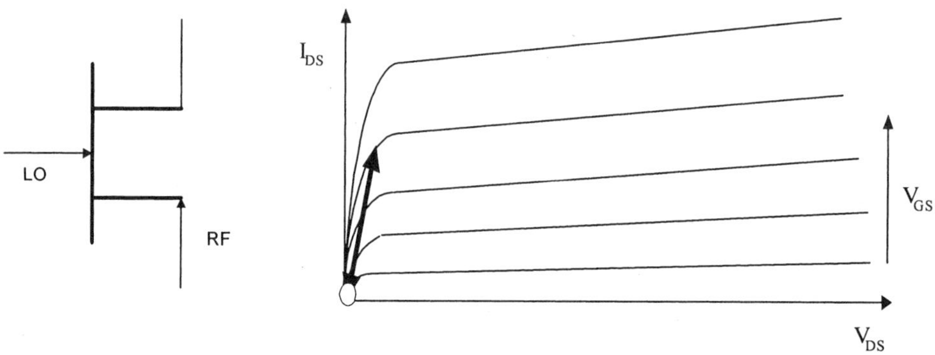

Figure 6-14. I-V Trajectory over LO cycle for TOPOLOGY #3a FET mixer.

5.2.3.3 TOPOLOGY #3b (alternate bias point)

In this instance, the LO and RF are fed to the FET in the same way as in TOPOLOGY #3a, but now the bias point is near V_{DD} and pinchoff, with the instantaneous bias trajectory shown in Fig. 6-15. This corresponds to the second bias point identified in the intuitive analysis. Again, the IF output is dependent on the LO component of the output conductance, which in this case is limited.

This biasing scheme is often recommended for mixer design. On studying the key dg_{DS}/dv_{GS} characteristic for this mixer configuration, however, it is clear that TOPOLOGY #3a is more suitable than TOPOLOGY #3b for good mixer performance.

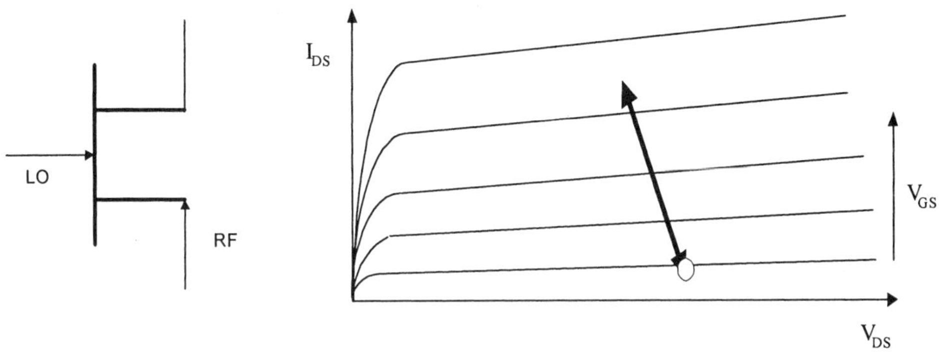

Figure 6-15. I-V Trajectory over LO cycle for TOPOLOGY #3b FET mixer.

Given the small-signal nature of the RF and the fact that, for this configuration, the IF current is dependent on the product of the RF voltage and the derivative of g_{DS} with respect to V_{GS} (i.e. the LO), it is apparent that the RF could equally well be applied to the drain terminal without impacting on the validity of the analysis. A LO-gate/RF-drain design would be expected to perform similarly to the TOPOLOGY #3 designs. This approach could be denoted TOPOLOGY #3 (alt).

TOPOLOGY #3 facilitates improved RF – LO isolation but, in this case, the LO leakage to the RF port can be a major concern. Variant (a) is preferable to (b) due to its intrinsically superior frequency conversion effects. However, (b) does offer the potential for some 'gain' which can redress this balance somewhat.

5.2.3.4 Practical Simulation Study

This topology was developed from the LO to source, RF to gate circuit (see TOPOLOGY #2), with the ports swapped and re-tuned. V_{DS} was set to 0V, corresponding to TOPOLOGY #3a. In this case, the predicted conversion performance for LO drive levels of +10 dBm and 0 dBm is summarised in Table 6-1.

Table 6-1. TOPOLOGY #3a Initial simulations of conversion loss for two LO power settings

RF (GHz)	Conv. Loss (dB) LO @ +10dBm	Conv. Loss (dB) LO @ 0dBm
57.2	6.0	13.5
58.2	6.2	13.8

A similar sensitivity of conversion loss to LO drive level is noted for TOPOLOGY #3a as was observed in the TOPOLOGY #2 simulations previously.

5.2.4 TOPOLOGY #4 LO and RF to gate (Transconductance mixer)

5.2.4.1 Generic Intuitive Study

In this configuration, both the LO and RF are applied to the gate. With the RF applied to the gate, the key FET parameter of interest, in terms of the IF generation, is g_m. This again follows from the definition of g_m, given that $v_{RF}(t)$ becomes associated with $v_{GS}(t)$. Similar to the analysis for the TOPOLOGY #1 approach, it follows that the optimum bias point is where the change in g_m with respect to the LO (and hence, in this case, with respect to V_{GS}) is maximised.

Consider Fig. 6-16. As indicated in the Figure, the solid lines correspond to a low value of V_{DS}, while the dashed lines correspond to a higher V_{DS}. It is clear that the rate of change of g_m with respect to V_{GS} is maximised for V_{GS} biased between pinchoff and the voltage where g_m peaks. It is also desirable that V_{DS} be biased in the saturation region for maximum g_m (and its derivative). The LO and RF are combined externally and applied to the gate. The instantaneous bias trajectory is shown in Fig. 6-17. The bias point is quite similar to that in TOPOLOGY #2 above. The optimum bias point is in the saturation region where the derivative of g_m with respect to V_{GS} is maximised.

TOPOLOGY #4 is the most commonly used FET mixer configuration. The Class A bias point means that some ‘gain’ is available to enhance the conversion effects. The disadvantage is that the LO and RF must be combined externally prior to injection at the gate. However, in order to achieve reasonable port isolation with any of the topologies above, some form of balancing is usually necessary anyhow. In this context, the RF and LO combination for TOPOLOGY #4 can be provided by a balancing structure (hybrid).

5.2.4.2 Practical Simulation Study

Again, for the purposes of initial simulation, a self-bias scheme was considered. The source resistance requires bypassing to ensure a good RF/LO ground and minimal RF power dissipation in the resistance. A simulation of the crude simplified schematic in Fig. 6-18 yielded the conversion performance presented in Fig. 6-19.

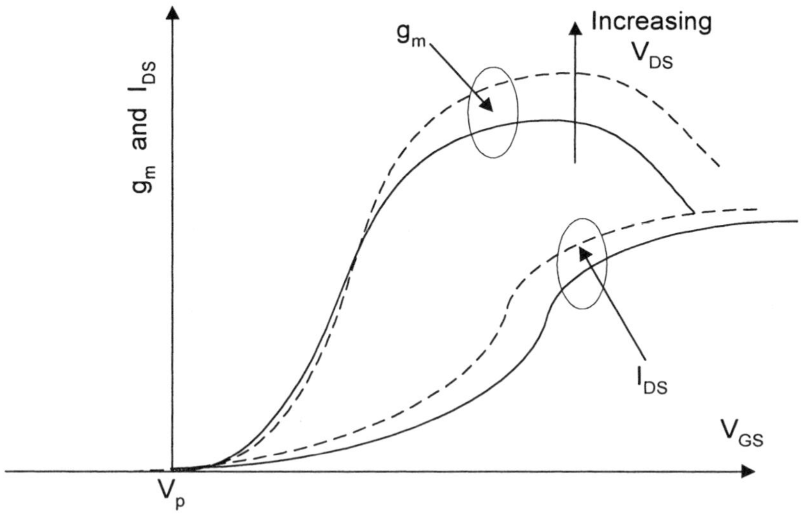

Figure 6-16. Representation of g_m and I_{DS} versus V_{GS} for a FET.

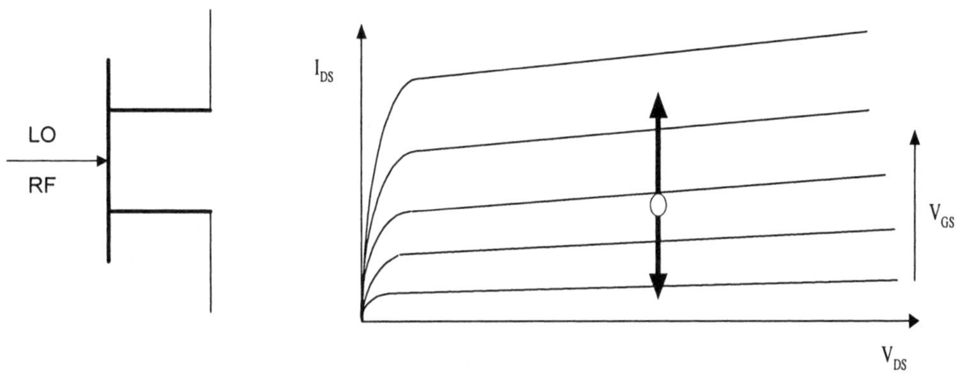

Figure 6-17. I-V Trajectory over LO cycle for TOPOLOGY #4 FET mixer.

As before, the input stub provides some LO/RF matching, while at the same time presents a low impedance to the IF when C_2 is suitably large. R_G is suitably large so that it does not load the input transmission line (not shown), but at the same time provides a zero potential at the gate when gate leakage currents are neglected. For the purposes of this simulation, the LO was fixed at 56.2 GHz, with a level of +10 dBm.

Note that negative conversion loss values here correspond to actual conversion *gain*. It is immediately clear from an initial consideration of this topology that it may offer superior conversion performance to the

TOPOLOGY #2 approach. The variation in the performance across such a narrow band could be a concern, but it should be remembered that the indicated performance is only an approximation at this stage.

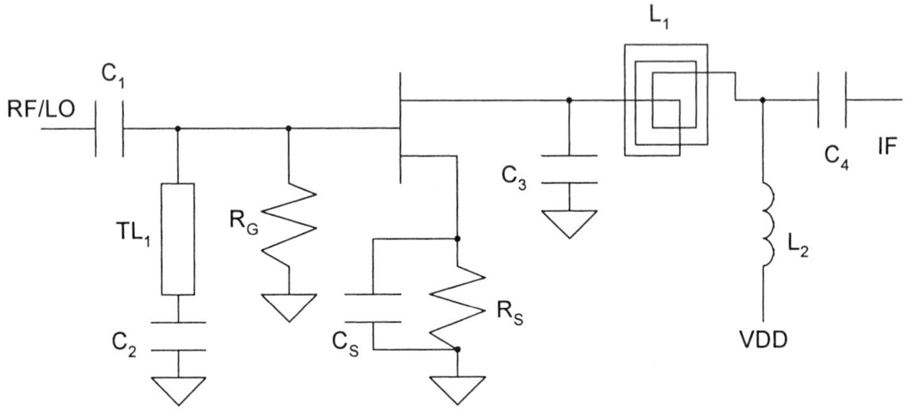

Figure 6-18. Initial simplified schematic for TOPOLOGY #4 FET mixer for simulation.

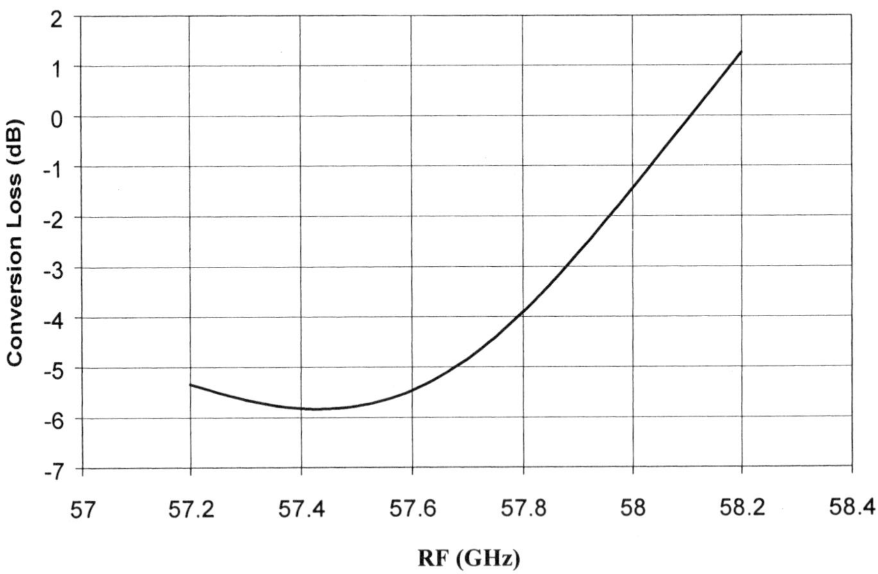

Figure 6-19. Simulated conversion loss for circuit in Fig. 6-18, for LO of +10 dBm at 56.2 GHz.

The same circuit was re-simulated, but with the LO level reduced to 0 dBm. The resulting conversion loss predictions dropped dramatically to about 20 dB.

5.2.5 General Comparison - Intuitive

As has been outlined above, each configuration has its own advantages and disadvantages from an intuitive standpoint, and some of the key points are summarised below.

TOPOLOGY #1 provides a convenient separation of the LO and RF inputs. The LO – RF isolation is enhanced by the reverse isolation of the FET device. Good RF – LO isolation may be difficult to achieve, and this could be an issue in some applications where frequency pulling of the LO source could result.

TOPOLOGY #2 again will manifest good LO – RF isolation. However, in this instance, the injection of the LO at the source can conflict with providing a good high frequency ground at that terminal. Moreover, due to the more complex nature of the LO modulation on the FET characteristics, the accurate simulation of its performance is likely to be a concern.

TOPOLOGY #3 on the other hand facilitates improved RF – LO isolation but, in this case, the LO leakage to the RF port can be a major concern. Variant (a) is preferable to (b) due to its inherently better frequency conversion properties. However, (b) does offer the potential for some 'gain' which can redress this balance somewhat.

TOPOLOGY #4 is the most commonly used FET mixer configuration. The Class A bias point means that some 'gain' is available to enhance the conversion effects. The disadvantage is that the LO and RF must be combined externally prior to injection at the gate. However, in order to achieve reasonable port isolations with any of the topologies above, some form of balancing is generally necessary.

5.2.6 General Comparison – 57 GHz Study Options

The various conversion losses for the different topologies as a function of LO drive are summarised in Table 6-2.

It is clear that the choice of optimum solution is dependent on the LO level available. Typically, LO levels of +10 to +13 dBm are used to pump mm-wave waveguide mixers. Such power levels are compatible with conventional LO sources (e.g. Gunn oscillators) and also with monolithic sources (see for example Chapter 7 on Frequency Multipliers). In view of this fact, it is apparent from Table 6-2 that the TOPOLOGY #4 FET mixer is the optimum topology from the perspective of conversion loss performance.

Table 6-2. Simulated FET mixer performance as a function of topology and LO level

LO Power (dBm)	RF (GHz)	Conv. Loss (dB) LO to gate, RF to source mixer TOPOLOGY #2	Conv. Loss (dB) LO to source, RF to gate mixer TOPOLOGY #3a	Conv. Loss (dB) LO/RF to gate mixer TOPOLOGY #4
+10	57.2	5.79	6.0	-5.34
+10	58.2	4.64	6.2	1.26
0	57.2	12.26	13.5	≅20
0	58.2	11.18	13.8	≅20

The remainder of this FET mixer simulation study focuses on the TOPOLOGY #4 approach.

5.2.7 Initial Parasitic Foundry Element Incorporation

The next key step in the mm-wave FET mixer development process was to take the basic circuit schematic above and to take account of the main parasitic effects by the inclusion of additional elements in the simulation. To a major extent, this is accomplished by using the specific foundry elements provided, in this case, by GMMT.

The parasitic elements involved include, for example, microstrip junctions such as tees and crosses, and GaAs via holes to the backside ground plane. The incorporation of the tees is critical as their parasitic phase characteristics offset the effective electrical lengths of associated transmission lines and stubs, and hence play an important role in accurately simulating the frequency response characteristic of a circuit. The via hole model factors in the effective inductance of this physical path to true ground, and this inductance can impact on circuit performance, particularly at high frequencies.

On including, to a first order, the primary parasitic effects, the circuit response is significantly degraded. The circuit elements were then re-tuned and the predicted response is presented in Fig. 6-20.

The original simulated performance from Fig. 6-19 is also included in Fig. 6-20 for convenience, and it is clear that the inclusion of additional parasitic effects has degraded the simulated conversion performance of the mixer. In essence, as the circuit being simulated is now made more 'real', the achievable performance is degraded. Moreover, the circuit requires re-tuning in order that the achievable (but degraded) performance be realised. This is highly significant, and emphasises strongly the absolute need to have good library models for the foundry process. These models must encompass not alone the conventional active and passive circuit elements, but must also accurately represent non-ideal parasitic components.

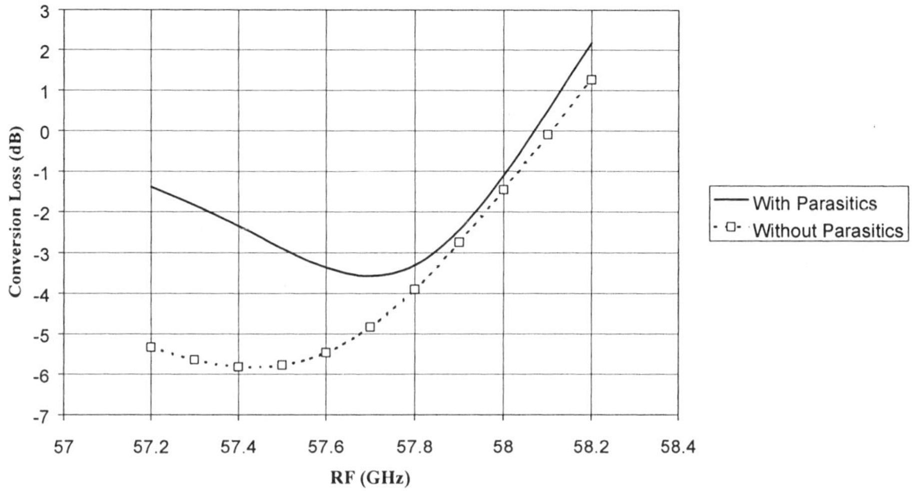

Figure 6-20. Impact of parasitics on TOPOLOGY #4 mixer conversion loss, for LO of +10 dBm at 56.2 GHz.

5.3 Detailed FET Mixer Schematic Development

At this stage, the FET mixer circuit topologies have been studied, and some interesting possibilities uncovered. The next step involves transferring those conceptual designs to layout-ready schematics. In fact, the layout of mm-wave circuits is a key component of the design, and the design iterates between schematic/simulation/layout many times before a truly optimised circuit is arrived at. As well as introducing the remaining foundry parasitic models, this step also includes bias network development and inclusion, and also balanced mixer studies for improved performance realisation.

5.3.1 Bias Network development and inclusion

Bias networks play a critical role in MMIC design. Because most of the key biasing components are on die, it is crucial that they are correctly designed; correction of on-die bias circuit errors after fabrication is generally impractical. At relatively low frequencies, in particular in RFICs, full temperature and process compensating bias controllers are generally implemented on the GaAs die. However, at mm-wave frequencies this is not usually the case, firstly because it would demand GaAs real estate, which is expensive for mm-wave material and, secondly, because for these high frequency GaAs processes, process control is not as mature as it is for lower specification processes. Thus, post-fabrication bias tunability *off-die* is prudent, particularly in development work.

For active circuits, the bias network is often used to enhance circuit stability by the incorporation of an out of band resistive path which tends to improve the immunity to bias oscillations. The challenge is to achieve this without compromising the in-band response.

For the purposes of the mixer developments, bias networks for the active FET mixers require more careful attention than those for diode-based designs due to the potential for oscillations in the active circuit. The careful selection of suitable decoupling capacitor values plays a crucial role in good mm-wave mixer bias network design. In particular, the requirement for physically small mm-wave decoupling capacitors, which do not introduce significant phase effects due to their transmission line properties, must be balanced against the requirement for larger capacitances when low frequency decoupling is involved. The challenge is to achieve this balance without compromising the in-band response.

Consider the network shown in Fig. 6-21

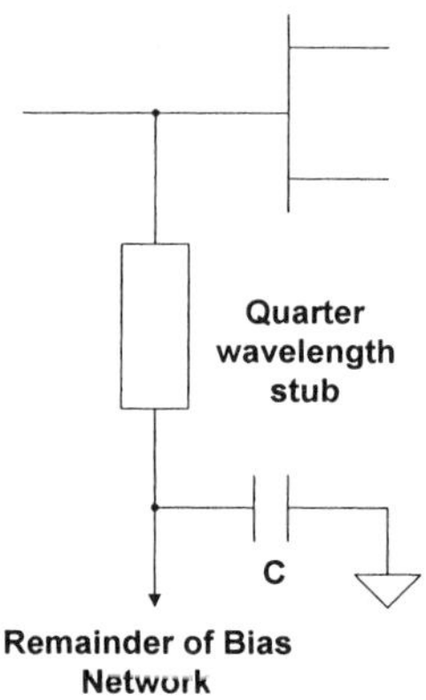

Figure 6-21. FET bias option 1.

In this arrangement, the capacitor presents a short circuit to ground at the end of the stub at high frequencies. This decouples the entire bias circuit from the main RF line at the FET's gate.

A difficulty arises in practice due to the non-ideal nature of the capacitor. In reality, this is a parallel plate component and is not an ideal lumped element value but has a phase length that becomes more significant at higher frequencies. As a consequence, the exact location of the short circuit plane is difficult to pinpoint and hence the effective length of the stub becomes somewhat unclear. Moreover, because the short circuit plane does not exist directly at the end of the stub, the impact of the remainder of the bias network is not fully decoupled from the main RF line. The tradeoff here is between, on the one hand a large capacitor to provide a better RF short-circuit at the expense of a larger phase length effect due to the capacitor and hence poorer decoupling of the remainder of the bias network, and on the other hand a smaller capacitor which has a reduced phase impact and hence results in more ideal decoupling of the remainder of the bias network, but at the expense of a poorer short-circuit due to the reduced capacitance.

An alternative approach which provides better decoupling of the bias network is shown in Fig. 6-22. This shows an interesting use of a spiral inductor in a mm-wave circuit. One of the issues associated with a spiral inductor is its self-resonant frequency. Above this frequency, the inductor stops behaving like an inductor and acquires capacitive and increasingly

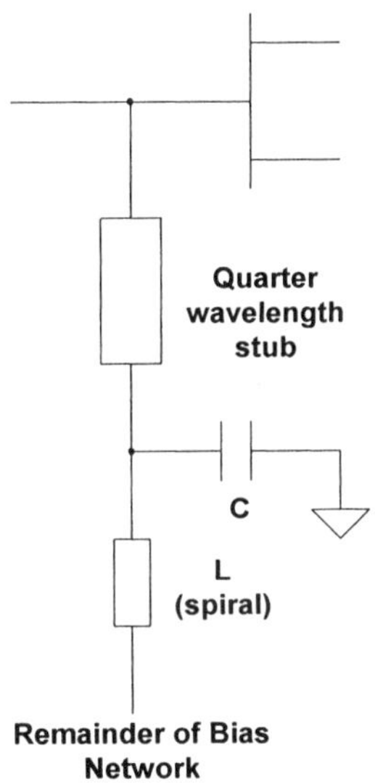

Figure 6-22. Alternative bias network.

complex characteristics. The self-resonance is due to the parasitics associated with the physical spiral, including such effects as inter-winding coupling capacitances. For typical spirals in the H40 process, the self-resonant frequency is in the range 10 – 20 GHz, and would certainly be well below the frequency range of interest for a 57 GHz transceiver application. In the bias scheme above, the small capacitor to ground at the foot of the stub decouples the mm-wave frequency signals effectively; excessive capacitor size can be avoided because the decoupling is being provided for high frequencies only. The series spiral inductor provides the lower frequency decoupling. The decoupling capacitor prevents the spiral from loading the mm-wave signals in an unpredictable fashion.

This latter bias scheme has been adopted extensively in the mixer circuits developed in the course of this work.

5.3.2 Balanced Mixer Considerations

So far, only single-ended FET mixers have been considered. The single-ended FET mixer has a number of interesting characteristics that have driven the research in this area for many years. However, the circuits have issues similar to those of single-ended diode mixers, from the standpoint of spurious response and isolations in particular. As a consequence, balanced designs are generally required in order that useful *system* level performance can be achieved.

Balanced approaches have been studied in detail and the optimum configuration for the 57 GHz transceiver development has been identified. An implementation of this approach using a novel mm-wave ratrace circuit has been developed and fabricated. Details on the balanced circuit analysis involved will now be presented.

A balanced mixer consists of a pair of single-ended designs connected between a pair of microwave hybrids. By careful design, the circuit can be arranged such that the overall conversion performance (ideal) matches that of the individual mixers while, at the same time, vastly improved return losses or isolations can be achieved. By suitable use of the hybrid characteristics, some of the unwanted mixing products can also be suppressed. A generic block diagram for a balanced mixer is shown in Fig. 6-23.

During the course of this work, the balanced mixer concept was studied in particular for the transconductance mixer topology. This mixer arrangement was selected due to the findings presented earlier where the conversion performance of this mixer was superior to that of the other topologies investigated. In Fig. 6-23, the LO and RF are combined in the input hybrid and then applied together to the 'gate' ports of the two single-ended mixers. The mixer outputs (drains) are then combined in the output hybrid and the IF is extracted from the appropriate port.

There are two forms of microwave hybrid. These are denoted 90° and 180° types. These hybrids are very different in how they process the input signals. Microwave hybrids have been described in great detail elsewhere[6] and only a synopsis of the important characteristics will be included here.

A generic block diagram of a 90° hybrid is shown in Fig. 6-24.

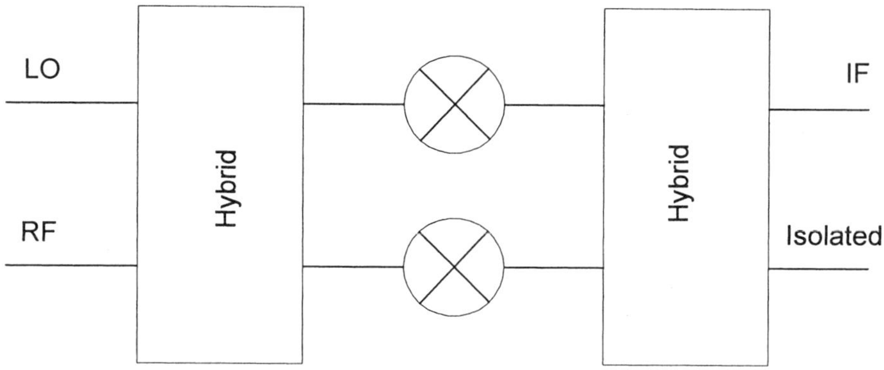

Figure 6-23. Generic balanced mixer block diagram.

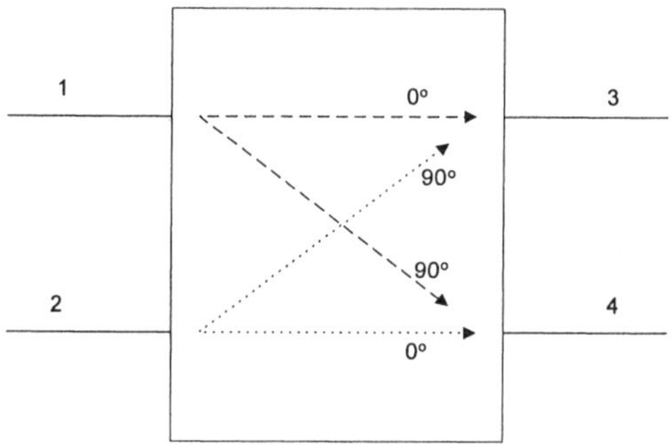

Figure 6-24. 90° Hybrid block diagram.

The hybrid is a four port element, and its transmission characteristics are explained as follows. A signal applied at port 1 is split equally and half the power is directed to ports 3 and 4, with port 2 being isolated. The relative phase difference between the signals emerging at ports 3 and 4 is 90°. A similar situation arises for a signal applied to port 2. In this case, the isolated port is port 1. The 90° hybrid is also referred to as a quadrature hybrid. A number of real implementations of this hybrid have been developed over the years; planar examples include the branchline and Lange couplers[6].

The 180° hybrid is shown in generic block diagram format in Fig. 6-25. A signal applied to port 1 is again split equally between ports 3 and 4, but in this in case, signals emerge with the same phase at ports 3 and 4. However, a signal applied to port 2 splits between ports 3 and 4 with a 180° phase difference.

Now, consider a balanced mixer consisting of various combinations of 90° and 180° hybrids encompassing a pair of identical single-ended mixers. This situation is shown in the 'arbitrary' block diagram in Fig. 6-26.

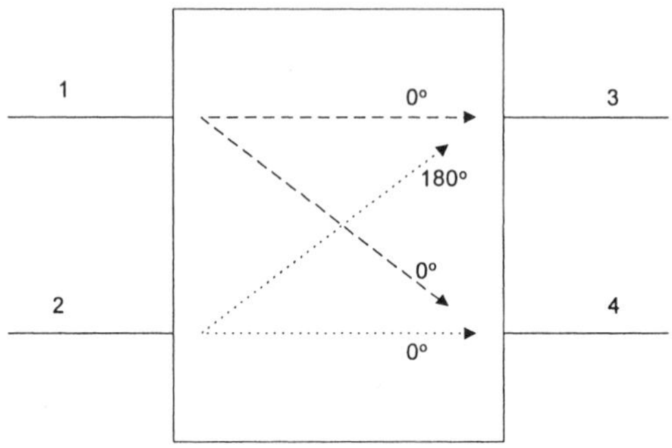

Figure 6-25. 180° Hybrid block diagram.

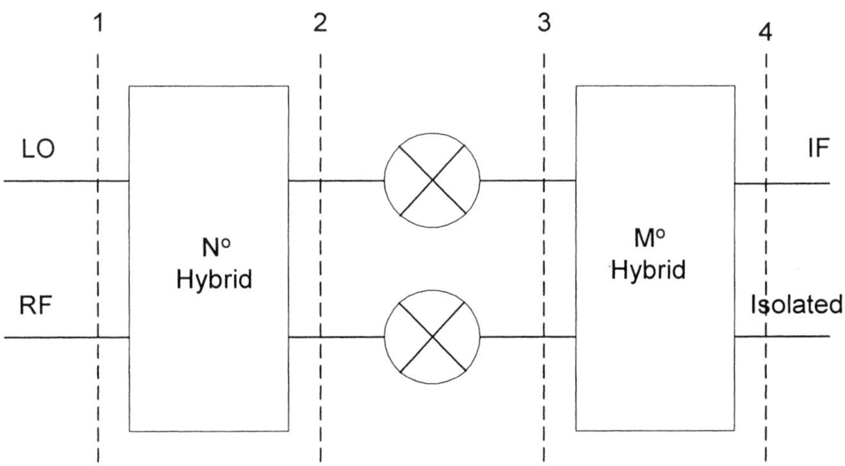

Figure 6-26. Balanced mixer with arbitrary input and output hybrids.

The values N and M can be either 90 or 180. Let us consider what the optimum choices may be for these values. In Tables 6-3 through 6-6, planes 1 – 4 correspond to the planes indicated in Fig. 6-26. Arms 1 and 2 correspond to the upper and lower arms respectively through the balanced structure (i.e. at all four planes). In the tables, the key frequency/phase components present on each arm at the four planes are shown. Only the IF component on the mixer output side is considered. Without any loss of generality, the RF is assumed to be above the LO frequency and the IF frequency/phase is derived from $f_{IF} = f_{RF} - f_{LO}$. The amplitudes are not shown as these assume equal values for both arms at each plane, and need not be considered at this point.

Table 6-3. Case 1: N = M = 90°

Arm #	Plane 1	Plane 2	Plane 3	Plane 4	Comment
1	$\omega_{LO}t$	$\omega_{LO}t$, $\omega_{RF}t + 90°$	$\omega_{IF}t + 90°$	$\omega_{IF}t + 90°$, $\omega_{IF}t$	In quadrature
2	$\omega_{RF}t$	$\omega_{LO}t + 90°$, $\omega_{RF}t$	$\omega_{IF}t - 90°$	$\omega_{IF}t + 180°$, $\omega_{IF}t - 90°$	In quadrature

Table 6-4. Case 2: N = 90°, M = 180°

Arm #	Plane 1	Plane 2	Plane 3	Plane 4	Comment
1	$\omega_{LO}t$	$\omega_{LO}t$, $\omega_{RF}t + 90°$	$\omega_{IF}t + 90°$	$\omega_{IF}t + 90°$, $\omega_{IF}t + 90°$	In phase
2	$\omega_{RF}t$	$\omega_{LO}t + 90°$, $\omega_{RF}t$	$\omega_{IF}t - 90°$	$\omega_{IF}t + 90°$, $\omega_{IF}t - 90°$	Cancel

Table 6-5. Case 3: N = 180°, M = 90°

Arm #	Plane 1	Plane 2	Plane 3	Plane 4	Comment
1	$\omega_{LO}t$	$\omega_{LO}t$, $\omega_{RF}t + 180°$	$\omega_{IF}t + 180°$	$\omega_{IF}t + 180°$, $\omega_{IF}t + 90°$	In quadrature
2	$\omega_{RF}t$	$\omega_{LO}t$ $\omega_{RF}t$	$\omega_{IF}t$	$\omega_{IF}t + 270°$, $\omega_{IF}t$	In quadrature

Table 6-6. Case 4: N = 180°, M = 180°

Arm #	Plane 1	Plane 2	Plane 3	Plane 4	Comment
1	$\omega_{LO}t$	$\omega_{LO}t$, $\omega_{RF}t + 180°$	$\omega_{IF}t + 180°$	$\omega_{IF}t + 180°$, $\omega_{IF}t + 180°$	In phase
2	$\omega_{RF}t$	$\omega_{LO}t$ $\omega_{RF}t$	$\omega_{IF}t$	$\omega_{IF}t + 180°$, $\omega_{IF}t$	Cancel

An interesting conclusion that can be drawn directly is that regardless of which type of hybrid is used at the input, the output hybrid should be a 180° type to ensure suitable combination and cancellation at the two output ports. In other words, Cases 1 and 3 do not merit further consideration.

The next question that needs to be considered is how to select the optimum type of input hybrid. This question is answered by considering the input return loss and isolation characteristics associated with Cases 2 and 4.

Firstly, consider the return loss characteristics. In case 2, with the input 90° hybrid, the reflections to the LO port from the two single-ended mixers cancel (provided of course that the mixing elements are identical). Similarly, the reflections at the RF port cancel. On the other hand, for case 4 (the input 180° hybrid), the overall LO and RF port reflection coefficients are associated with the corresponding reflections at the single-ended mixer ports. As a consequence, where port return loss is of prime importance, the case 2 style implementation is optimal.

Now, consider the LO to RF isolation. For case 2, the LO signal coupled to the RF port consists of the LO reflections at the single-ended mixers directed back to the RF port. It is easy to see that these signals add in phase. Thus, the LO-RF and RF-LO isolation are the same as those of the individual single-ended mixers. On the other hand, for case 4, with the input 180° hybrid, these L-R and R-L isolations are theoretically perfect as the two reflected signals are out of phase and cancel. As a consequence, where the isolations are of particular importance, the case 4 configuration is optimal.

In the 57 GHz transceiver application of interest to this work, the RF and LO return losses of the single-ended mixers can be designed to be at reasonably high levels, and consequently we deemed the case 4 configuration (with input 180° hybrid) the most appropriate in order that good RF-LO isolations be achieved. As will be shown later, the RF and LO matches can be further improved if necessary.

The most commonly employed planar 180° hybrid is the ratrace[6]. In this work, a reduced size ratrace[64] was employed to form the input hybrid. To the authors' knowledge, this is the first published use of this approach in an MMIC development. The reduced size ratrace has a number of associated benefits including reduced die size and reduced ohmic losses in the hybrid itself.

Due to the low IF (and hence long wavelength), the most suitable 180° IF hybrid is some form of off-chip transformer based structure. To implement such a hybrid on GaAs would use excessive GaAs real-estate. As a consequence, the balanced FET mixer design developed during the course of this work has been laid out with a balanced pair of IF outputs. Both IF ports were laid out with probe pads such that on-wafer probing to the individual IF ports was facilitated.

Of course, as two separate single-ended mixing elements are involved, the LO drive requirement of the balanced circuit it typically a little more than 3 dB greater than the single-ended mixer drive requirement. This has clear implications in terms of mm-wave power generation for the LO source. In the system level approach studied in this work, the mm-wave LO signal is derived from a frequency tripler driven by a lower frequency source. This low frequency signal is relatively cheap to generate, and moreover is more readily phase locked than a direct mm-wave source. The goal of the 57 GHz multiplier study, described in Chapter 7, was to provide adequate power at the third harmonic to pump optimally the two single-ended mixing elements in a balanced configuration.

5.3.2.1 Ratrace Hybrid

The ratrace is one of the oldest microwave hybrids, and is very commonly used in planar/MMIC applications. A representation of the upper conductor in a microstrip ratrace structure is shown in Fig. 6-27. In this conventional ratrace, the ring structure is 1.5λ in circumference (at the centre frequency).

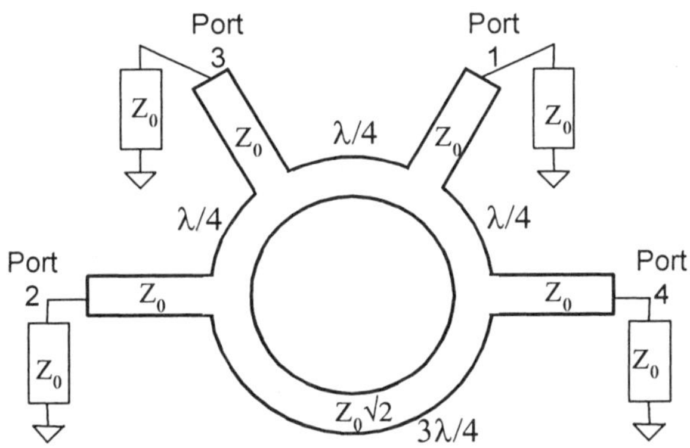

Figure 6-27. Conventional ratrace structure.

A detailed explanation of the operation of this structure is presented elsewhere[6], and will be only mentioned briefly here. Ports 1 and 2 are isolated. This can be understood very simply by realising that there are two parallel paths between ports 1 and 2, and these differ in length by half a wavelength. As a consequence, the equal signals that propagate from port 1 to 2 (and vice versa) cancel and isolation is thus achieved. Similarly, ports 3 and 4 are isolated. On the other hand, a signal applied to port 1 has two paths to port 3. The path length difference in this case is a full wavelength and the signals add in phase. The same applies to the coupling from port 1 to port 4. Ports 3 and 4 are the same distance from port 1, and thus the signals emanating from ports 3 and 4 are in phase.

On the other hand, a signal applied to port 2 also couples to both ports 3 and 4, but in this case, the signals emanating from the latter two ports are out of phase due to the asymmetry of the ratrace structure. This gives rise to the 180° hybrid characteristics. The non-idealities of the ratrace are primarily due to ohmic losses in the metal structure.

The ratrace structure has a bandwidth of about 10 – 15%. For the intended 57 GHz transceiver application, this would not be an issue, and in fact exceeds the requirement. An alternative ratrace structure has been proposed by Kim and Yang[64]. Here the conventional ratrace was modified and the overall circumference shortened. By appropriate selection of line impedance and length, useful hybrid characteristics were still obtained by Kim and Yang. The characteristics of the hybrid over frequency are different to those of the conventional structure, but for relatively narrow band applications are similar, and in fact are likely to be better due to lower ohmic losses. This modified structure is represented in Fig. 6-28[64].

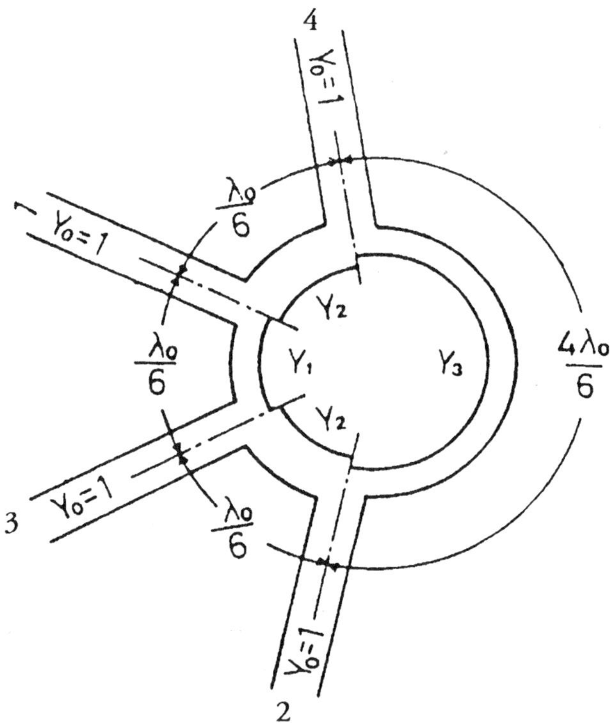

Figure 6-28. Reduced size ratrace structure (source Kim and Yang[64], © 2004 IEEE).

The total circumference of this structure is 7/6λ which is significantly less than that of the conventional design. Simulated s-parameters for a microstrip version of this approach using the H40[3] transmission line models are summarised in Fig. 6-29. In this plot, it is seen that the transmission responses from port 1 to ports 3 and 4 are both around -3 dB. The phase difference between these two signals is represented by *Del(phase1)* in Fig. 6-

29, and lies close to 0°. Also shown is the corresponding phase difference for the two signals at ports 3 and 4 due to a signal applied to port 2. This phase difference is represented by *Del(phase2)*, and it is clearly close to 180° across the range 55 – 60 GHz. The amplitude responses from port 2 to ports 3 and 4 have been omitted from the plot for ease of readability, but these are also close to –3 dB.

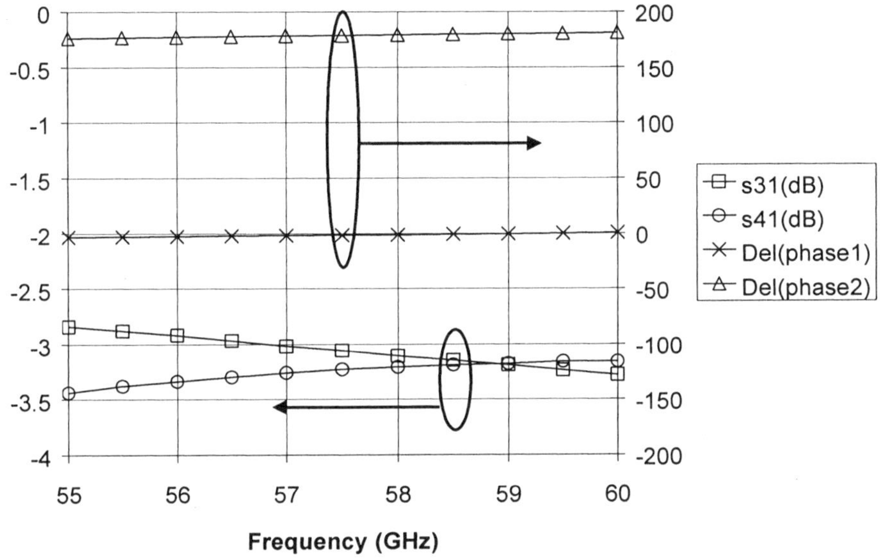

Figure 6-29. Simulated performance of reduced size retrace.

When a balanced mixer, with identical single-ended mixing elements, is built up as a schematic with the input LO/RF connections being made to the above ratrace and the output consisting of an ideal 180° hybrid, and the resulting circuit is simulated, the predicted conversion loss performance is outlined in Fig. 6-30. Also shown in this Figure is the corresponding predicted performance when this ratrace is replaced by the conventional design (also implemented with H40 transmission lines). It is immediately clear that the reduced size structure does offer the potential for improved conversion performance by virtue of reduced ohmic losses.

This is not the final mixer design, but serves as a useful means of comparing the relative performance of the two ratrace approaches. Given the clear indication of reduced ohmic losses in the reduced size option, this approach was adopted for the subsequent balanced mixer development work.

5.3.3 Up Converter Considerations

The design of an up converter has typically received much less attention in terms of design methodology in the literature than down converter design. This is unfortunate, as there are some aspects to up converter design which are not relevant to down converters, and vice versa.

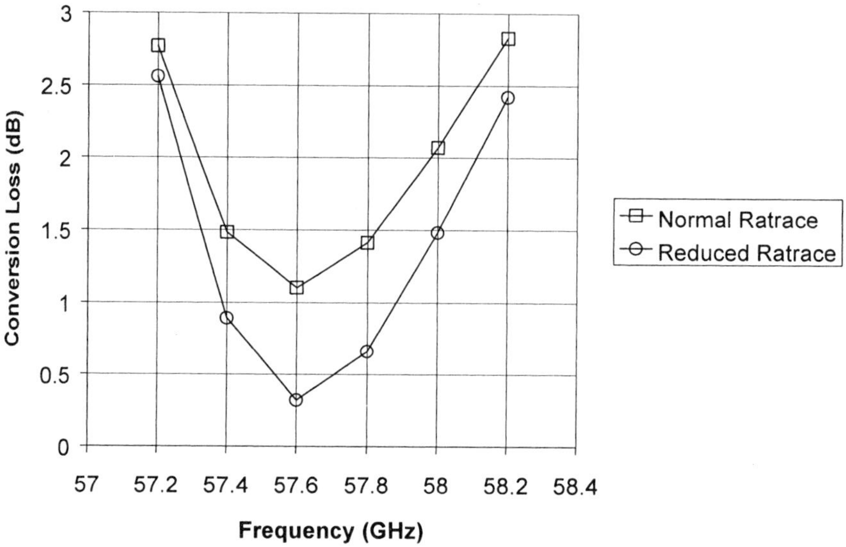

Figure 6-30. Simulated conversion loss of balanced mixer using each ratrace structure.

As in the case of a down converter, in a single-ended up converter, the LO and RF are often connected to different terminals of the FET. For instance, it is typical for the LO to be applied to the gate, whereas the RF is commonly extracted from the drain. IF injection is often at the drain. This typically means that good LO to RF isolation may be difficult to achieve, depending on the separation of the RF and LO frequencies. Balancing techniques are often employed to improve this situation. A possible scheme is outlined in Fig. 6-31.

In this configuration, the LO signals applied to the two individual mixers are 180° out of phase, by virtue of the hybrid characteristics. The IF signals at the single-ended mixers are similarly out of phase. As a consequence, the RF signals generated in the two mixers are in phase, and combine in a 0° combiner. On the other hand, the LO and IF signals leaking through the individual mixers remain out of phase. Thus, assuming that the RF combiner has a reasonable bandwidth, and that its 0° phase characteristic is approximately valid at the LO, then good cancellation of the LO at the RF port follows.

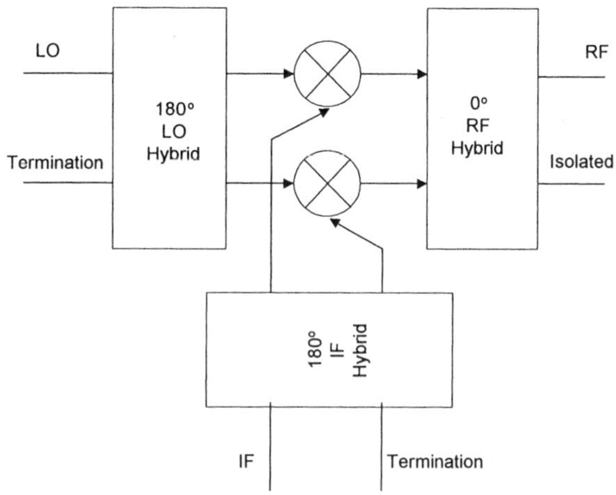

Figure 6-31. Possible balanced up converting mixer topology.

In terms of planar implementation, the LO hybrid can be implemented as a ratrace, similar to the structure outlined above for the down converter. The RF combiner could be implemented as a Wilkinson structure. The Wilkinson hybrid is in theory a 180° structure, with a virtual node embedded in its resistive termination. In practice, it behaves as a relatively broadband 0° combiner/splitter. As in the case of the down converter, the IF hybrid is typically implemented off chip, and for a 180° requirement, a wound transformer is often used.

A key issue with FET mixers is that, unlike a diode, the FET device is not reciprocal. In other words, in up converting mode, the RF cannot be extracted from the gate terminal. Since a FET down converter is often configured with the RF applied to the gate (the transconductance mixer for instance), such a mixer will not function very well as an up converter. This is one area where the diode mixer has a significant advantage.

5.4 Balanced FET Mixer Design Details

The schematic of the transconductance mixer, described earlier, was further developed to incorporate the various requirements discussed in Section 5.3 of this chapter. This circuit was then itself incorporated into a balanced configuration, using the reduced size ratrace hybrid. The final balanced FET mixer schematic is presented in Fig. 6-32. Firstly, the combination of two single-ended mixing circuits with the ratrace and LO and RF input networks can be seen. The two mixing circuits are identical. The LO and RF signals, combined by the ratrace, are fed to the device gate terminals. The network between the ratrace and each gate performs a number of functions:

- DC block
- LO/RF match into device
- Gate bias injection
- Relatively low IF impedance at gate by virtue of C3.

Spiral inductor L2 is used to further decouple external biasing elements from the circuit at low frequencies.
For flexibility, the FET gates can be biased (VGG1, VGG2), but a self-bias solution is much more preferable for a volume design, and such an arrangement is also facilitated by this design. The self-bias components are C3 and R1.

On the drain side, the LO and RF are terminated in a reactive impedance by virtue of the shunt capacitors, C1 and C2. These capacitors are small in value and are not very significant at the IF. These capacitors effectively decouple the high frequency signals from the circuit beyond, and therefore the spiral inductors L1 can be safely used for IF matching and bias injection purposes. If capacitors C1 and C2 were not present, L1 would have an unpredictable impact on the mm-wave signals at the drain.

The two IF ports, IF+ and IF-, are required to interface with an external 180° IF hybrid for combining purposes. The LO and RF networks include open stubs to facilitate improved matching to the ratrace. The RF, LO and IF ports are all implemented as ground – signal – ground coplanar structures that are suitable for wafer probing as well as wire bonding. The conventional probe configurations supported by GMMT, and for which models are available, are not exactly 50 Ω structures, and as a consequence some small amount of matching does help somewhat. However, the primary reason for the stubs is that the RF and LO matches of the individual single-ended mixing elements are not ideal and, as shown earlier, the hybrid configuration used was selected for isolation reasons, and the match characteristics of the balanced structure are the same as those of the single-ended circuits.

The layout of the final balanced FET mixer is shown in Fig. 6-33. Note that the ratrace hybrid structure in the centre consists of straight rather than circular or curved transmission lines; this is the case because the GMMT model set[3] only supported straight microstrip lines. Using specific bend elements, which are available in the model set, however, the parasitic effects associated with the corners were included in the simulation schematic and thus they were suitably compensated for.

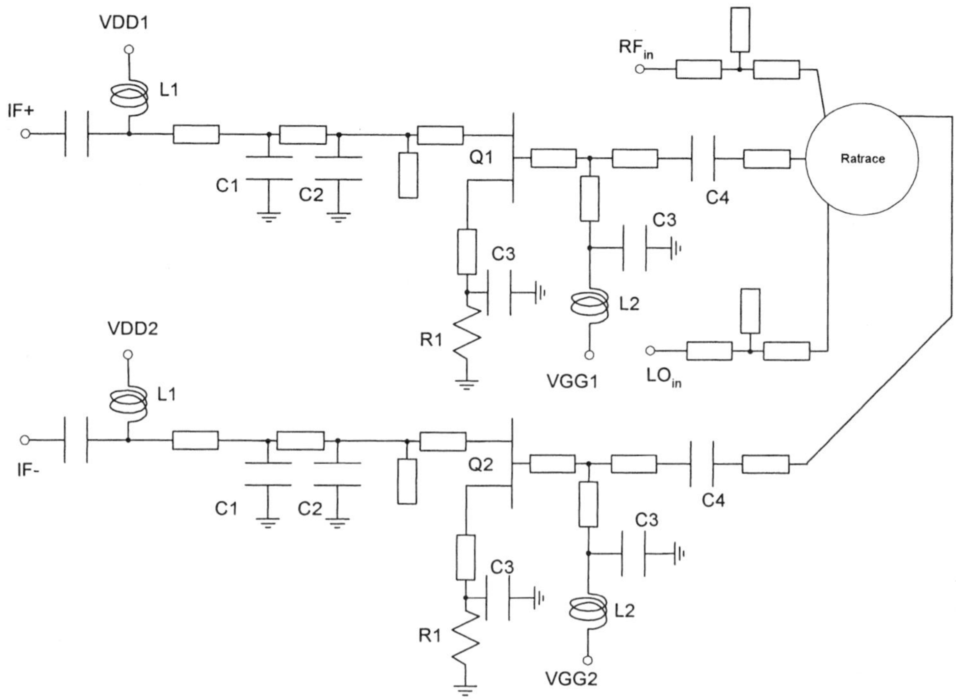

Figure 6-32. Final schematic of balanced FET mixer fabricated using H40 process.

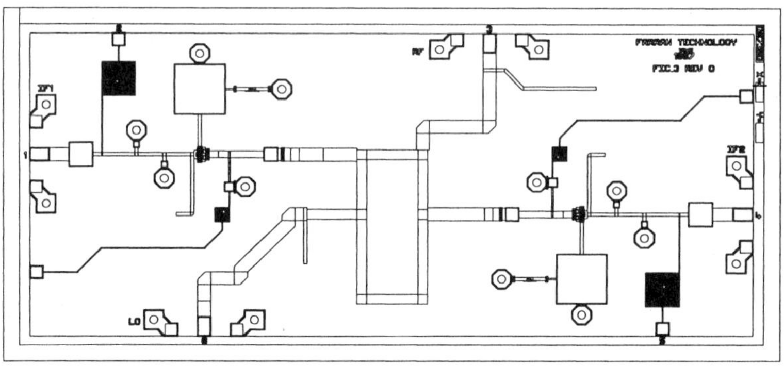

Figure 6-33. Balanced MMIC FET mixer layout.

The simulated conversion loss characteristic of this mixer is shown in Fig. 6-34. The circuit is simulated under the conditions:

- LO fixed at 56.2 GHz, +13 dBm
- RF swept, -30 dBm

This simulated performance is comparable to the conversion loss achievable with conventional waveguide mixers based on packaged Schottky diodes. The monolithic approach of course lends itself to a higher level of integration and this is one of the key drivers in the development of MMIC mm-wave transceivers. The conversion loss is reasonably flat with frequency, with a minimum being noted at an RF of 57.4 GHz.

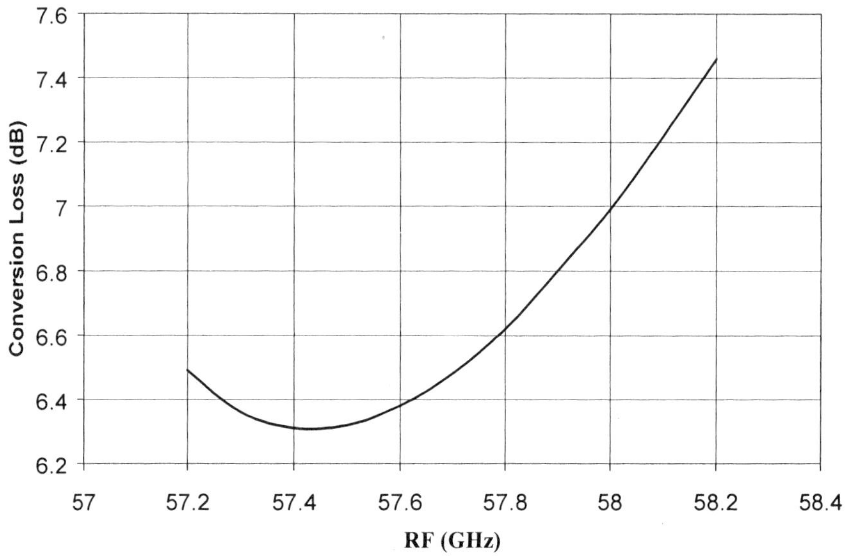

Figure 6-34. Simulated conversion loss for fabricated balanced FET mixer, for LO of +13 dBm at 56.2 GHz.

It is noted that this simulated performance is about 7 dB worse on average than that presented earlier in Fig. 6-20. This shows the level of degradation introduced when the bias networks are modelled and a schematic consistent with good layout practice is developed.

5.5 Schottky Diode Mixer Design Considerations

Schottky Diode mixers have been a topic or research for many years. Typically, these mixers have been implemented in waveguide-based configurations for high frequency applications. However, the same design concepts in terms of balancing, isolation optimisation etc. apply in waveguide and in MMIC technology. Hybrids are required in both cases. The structures of the hybrids are of course entirely different in the two technologies, but the underlying principles and the fundamental purpose of the hybrids remain unchanged.

The diode mixer is being replaced in many MMIC designs by a FET equivalent device. However, there are certain situations where a diode mixer still finds considerable favour. Generally, the performance of a diode mixer is better understood from a theoretical standpoint. Moreover, the bias/performance tradeoffs are much simpler for a diode mixer. For a given diode bias (and current), the performance of a diode mixer generally improves with increased LO level until saturated performance is obtained. This is not the case with the FET mixer where an optimum bias point exists for a given LO level.

The diode mixer also has the major advantage that it can simultaneously operate as both an up converting and down converting circuit. This is not necessarily the case with the FET approach due to the non-reciprocal nature of the FET characteristics.

A given diode mixer design can operate equally well in both up and down converting modes, and given the requirement for both mixer types in the 57 GHz transceiver, such a feature was of considerable interest in this work.

The same balancing considerations discussed above for the FET mixer development also apply for diode based circuits. The hybrid requirements can be addressed in a similar manner.

The balanced diode mixer has the key advantage over the FET approach that it does not require the output 180° IF hybrid. This is the case because, being a reciprocal two-terminal device, one of the diodes in the single-ended mixing circuits can be reversed. This has two desirable impacts. Firstly, the diodes can be series connected via a DC common node in the input ratrace hybrid, and may thereby be biased by a single DC supply. The resulting DC current is the same for both devices, and thus their bias points are automatically well matched. Secondly, the signals of the IF outputs are now in phase and thus instead of an output 180° hybrid being required, a direct connection becomes possible. This simplified output scheme is shown schematically in Fig. 6-35.

Due to this distinct packaging advantage, the diode mixer was implemented in both the up and down converting arms of the 57 GHz transceiver developed in the course of this work. No specific non-linear diode device or model was made available by the foundry. As a consequence, the pHEMT model was used for the diode mixer simulations – the diode was realised by short circuiting the source-drain terminals of the pHEMT device, and the same connections were made to the pHEMT model for simulation purposes. It was realised at the outset that the use of such a diode structure was not ideal for two reasons. Firstly, the use of the pHEMT model with the shorted source/drain makes the assumption not alone that the diode portion of the pHEMT model represents the actual Schottky diode well, but also that a model primarily extracted for negative gate - source bias voltages is valid for the forward bias levels used in diode mixers. Secondly, the optimum material layer structure for a Schottky diode, where these forward bias voltages typically apply when used in mixer design, is distinctly different to the structure most appropriate for low noise pHEMT device performance.

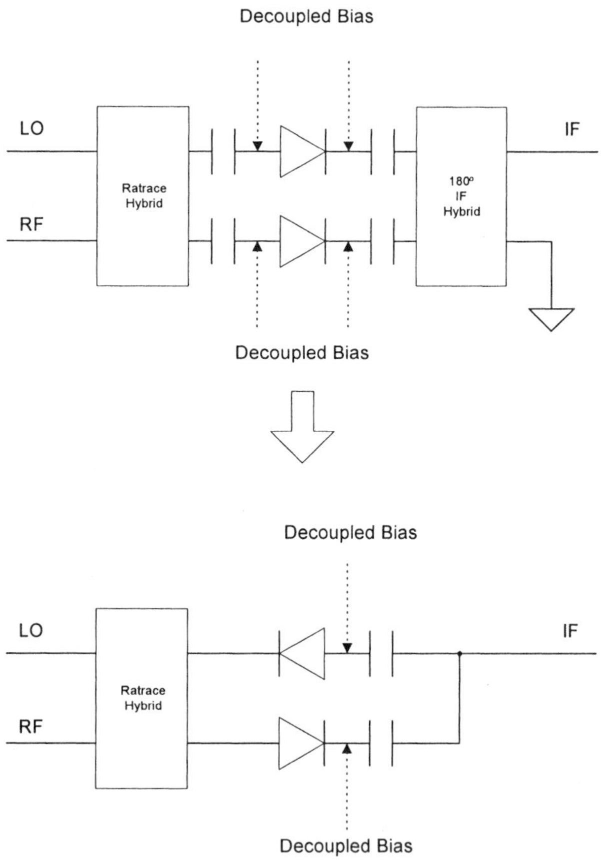

Figure 6-35. Elimination of output hybrid in balanced diode mixer.

Nevertheless, in the absence of any specific diode device in the foundry's standard offerings, given the potential advantages that a diode mixer has over its FET counterpart, it was considered prudent to pursue this mixer option with its possible shortcomings in mind.

5.6 Balanced Diode Mixer Design Details

The schematic of the balanced diode mixer design is shown in Fig. 6-36. This is very similar to the FET based design, but the biasing is simpler. The two diodes are implemented as pHEMT devices with their sources and drains shorted together. Moreover, only a single IF port is required due to the direct connection explained above. Note that, in this case, the RF and LO networks do not include open stubs for matching, as it was found during the design development that good matches could be achieved without them. Note also that the circuit was implemented with DC blocks on the individual single-ended mixing elements. This prevents the automatic biasing of the two diodes with the same current but, in return, introduces an additional degree of freedom that is prudent in a development of this type. This design has potential to operate as both a down-converter and an up-converter. The corresponding balanced diode mixer layout arrived at is presented in Fig. 6-37.

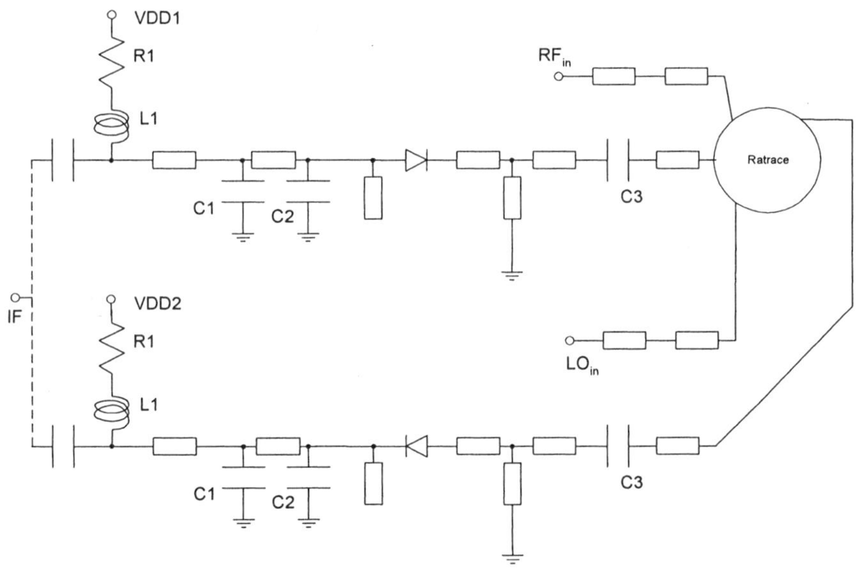

Figure 6-36. Final schematic of balanced diode mixer fabricated using H40 process.

Simulated conversion losses for these two modes of operation are shown in Fig. 6-38. Note that, in this figure, “U/C” denotes Up Converter and “D/C” denotes Down Converter. For these simulations, the LO has been fixed at 56.2 GHz at a level of +10 dBm, and the RF has been swept over the desired frequency range of 57.2 – 58.2 GHz at a level of –30 dBm. It is found in simulation that the conversion performance is relatively unchanged for LO drive levels in excess of +10 dBm.

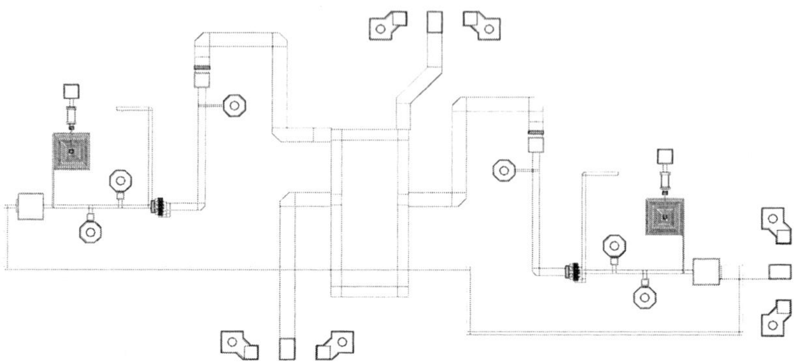

Figure 6-37. Balanced MMIC diode mixer layout.

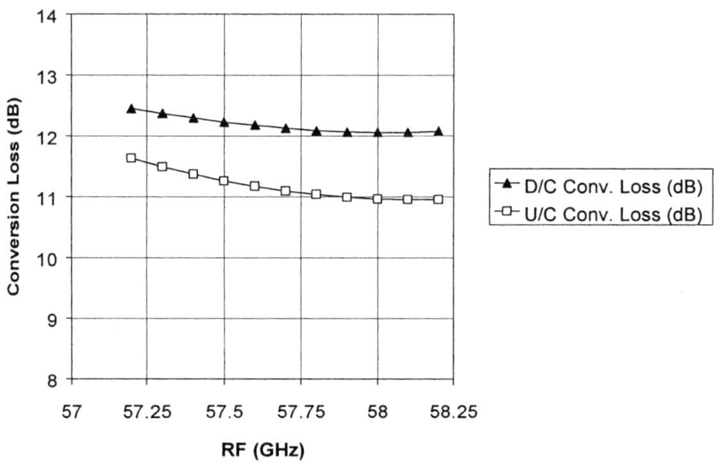

Figure 6-38. Simulated conversion loss for fabricated balanced diode mixer, with LO of +10 dBm at 56.2 GHz.

As expected, the design shows promise of useful operation in both modes, with similar conversion losses being predicted for both down and up conversion – the indicated up conversion loss is approximately 1 dB lower than the corresponding down conversion loss. It can be seen that the predicted conversion loss in the down-conversion mode is about 6 dB poorer than the predicted values for the corresponding balanced FET down-converter (Fig. 6-34). This immediately demonstrates one of the key benefits associated with the use of the FET in mixer development – being an active device, it introduces the potential for 'gain'.

The diode mixer's predicted conversion loss is also about 6 dB higher than that typically realised with a conventional waveguide diode mixer. It must be borne in mind that the Schottky diode used in this work is not an optimised physical diode structure, nor is the electrical model optimised for diode simulations.

5.7 MMIC Mixer Evaluations

The measured performance of the fabricated balanced FET and Schottky diode mixers is now described. A summary comparison of the measured and simulated conversion loss performance is provided in Fig. 6-39 as an aid to the reader. The mixer circuits were evaluated on wafer and also in customised test blocks.

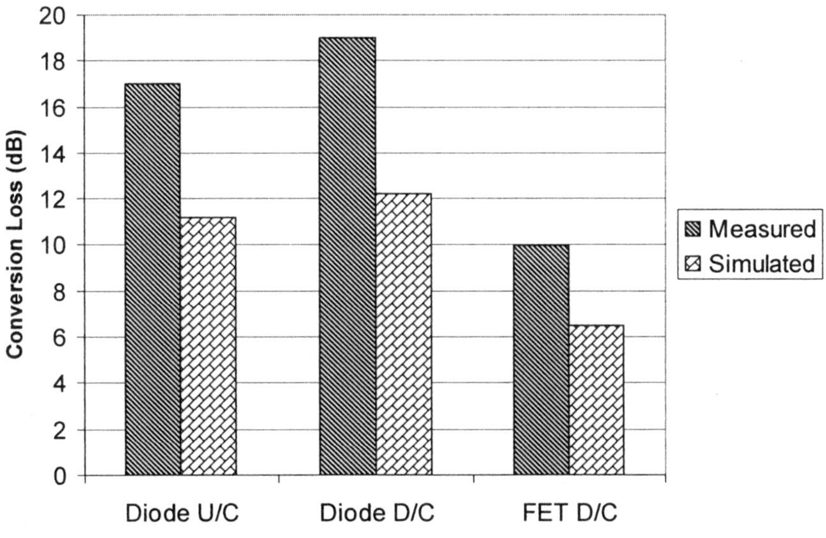

Figure 6-39. Comparison of measured and simulated mixer conversion loss for the balanced FET and diode mixers.

5.7.1 FET Mixer Evaluation

5.7.1.1 On-Wafer FET Mixer Characterisation

The on-wafer measurements were carried out in the University of Glasgow, using a high frequency probe station and associated high frequency test equipment. For the mm-wave signal interface to the wafer, V-band GSG Picoprobes, described in Chapter 2, Section 6, were used; these are specified to 67 GHz. The IF signal interface was provided by a lower frequency GSG probe. The bias connections were provided by needle probes. The decoupling of these needle probes was quite primitive, and in hindsight, unsatisfactory – see also Section 2.3 in chapter 5. It comprised of chip capacitors soldered between the top of the probe and RF ground, and was inadequate due to the long length of inductive needle between the tip and the decoupled point. This decoupling approach placed limitations on the biasing that could be applied to the active devices without bias oscillations being observed.

For the purposes of LO generation, a high frequency source was used in conjunction with a mm-wave power amplifier module – due to test equipment constraints, the available LO power for the on-wafer testing was +10 dBm instead of the simulated +13 dBm. For down converter measurements, the RF signal was provided by a Wiltron-360 Vector Network Analyser and the IF was routed to a Spectrum Analyser. The down converter test setup is shown in Fig. 6-40.

The conversion to one of the IF ports was characterised with the other port being terminated. The optimum performance was measured for the following conditions:

- LO input +10 dBm at 56.2 GHz
- RF input at 56.56 GHz
- V_{DD} 0.6V, V_{GG} –0.9V
- Conversion loss to each IF port 17 dB

If a suitable IF hybrid were fitted, an improvement of about 3dB would be expected in the conversion loss, yielding a predicted value of 14 dB. This is of the order of 7-8 dB poorer than simulated.

The bias point does not coincide with that expected. It was found during the on-wafer measurements that oscillatory behaviour was observed when V_{DD} was increased above approximately 0.6V. This oscillation was similar to that observed during the on-wafer LNA characterization, as described in Chapter 5.

5.7.1.2 Packaged FET Mixer Characterisation

The HEMT Mixer was packaged in a customised metal block with coaxial interfaces. The block was designed at Farran Technology Ltd. A drawing of the assembly is shown in Fig. 6-41. An IF hybrid was not incorporated in the body. Both the RF and LO ports were V-type female connectors; the IF connectors were SMA-female. 50 Ω microstrip lines on quartz formed the interface from the box edge to the RF and LO bond-wires (and then onto the die). The corresponding IF connections were fabricated using 50 Ω lines on duroid. With this arrangement, improved decoupling at the bias inputs becomes possible by suitable location of capacitors adjacent to the MMIC bias pads, on the base of the metal box.

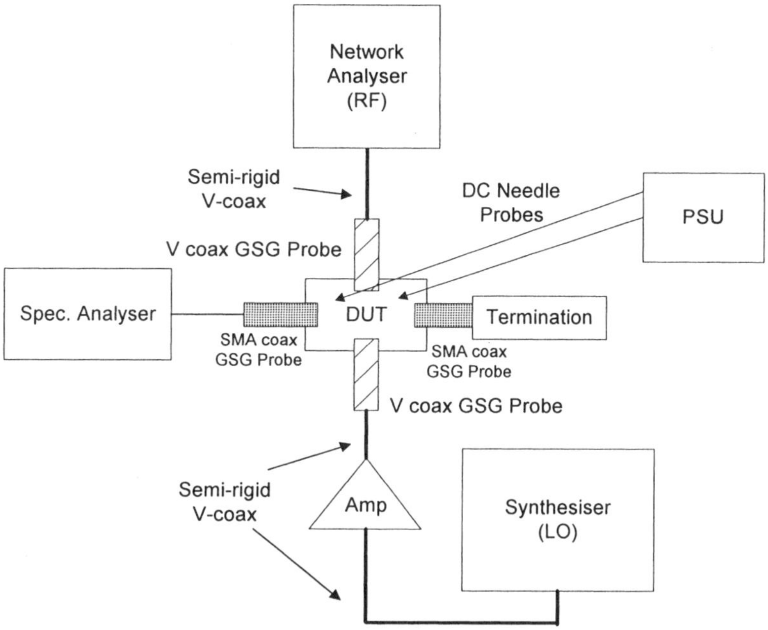

Figure 6-40. On-wafer test setup for down converting balanced FET mixer.

The block was machined and assembled at Farran Technology Ltd., and the testing was also carried out at that facility. An Anritsu-Wiltron 541771A Scalar Network Analyser and other mm-wave test equipment were used for this characterisation. The test arrangement is presented in Fig. 6-42.

With the Gunn oscillator based LO source, it was possible to deliver the desired +13 dBm LO drive for the packaged evaluations. In this case, the measured conversion loss to each IF was 13 dB. If a suitable IF hybrid were fitted, an overall conversion loss of 10 dB would be expected. This 4 dB improvement in conversion performance relative to the on-wafer measurement is primarily attributed to the increased LO drive level available for the packaged test. The 10 dB conversion loss is about 3 dB worse than simulated. The cause of this discrepancy is mainly attributed to inadequate model accuracy in the frequency range of interest.

Nevertheless, the FET mixer was certainly functioning optimally at the target mm-wave frequency, suggesting that the passive elements of the circuitry were well tuned. The poorer than predicted performance will be discussed further in Section 6 of this chapter.

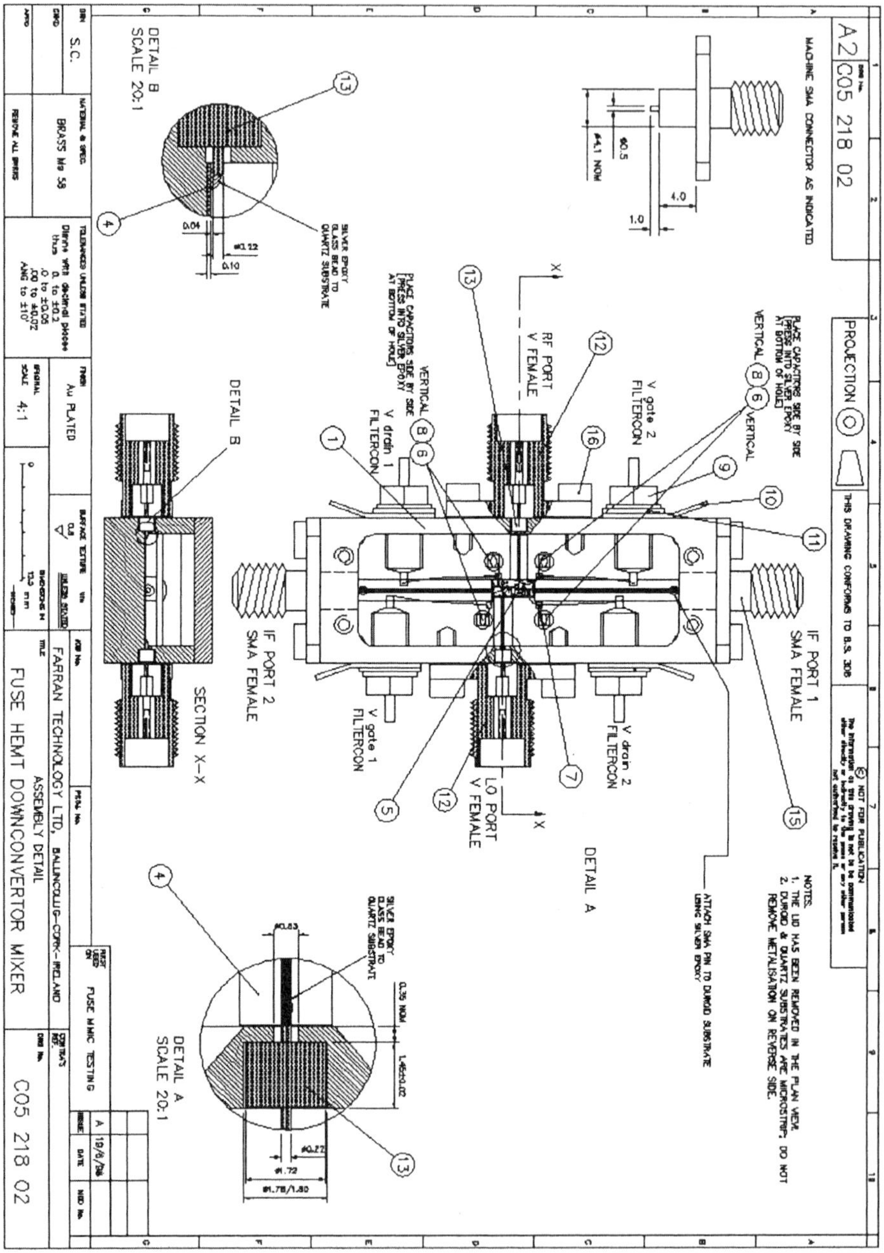

Figure 6-41. Assembly drawing for balanced FET MMIC mixer.

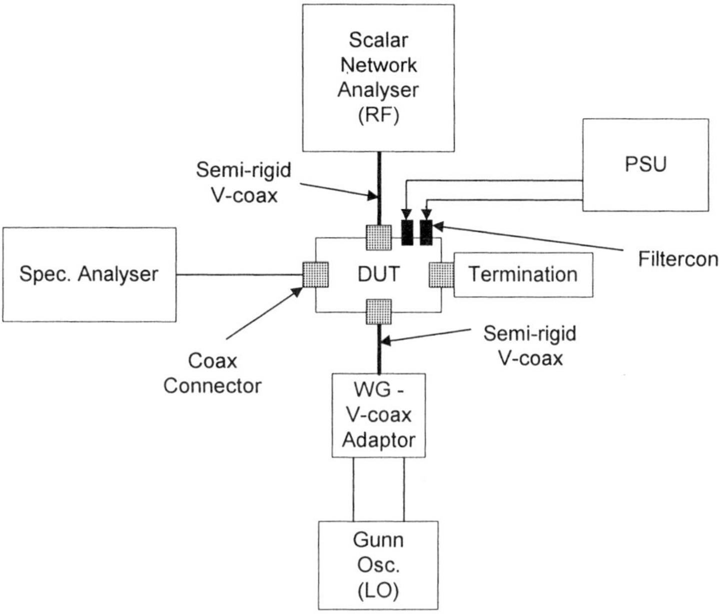

Figure 6-42. Test setup for packaged down converting balanced FET mixer.

5.7.2 Diode Mixer Evaluation

5.7.2.1 On-Wafer Diode Mixer Characterisation

The measurements were carried out using the same test setup described for the FET mixer, and depicted in Fig. 6-40, with the key distinction being that only a single IF connection is involved in the case of the diode mixer. The optimum performance realised is summarised below for both modes of operation.

Down-converter mode:

- LO +10 dBm at 57.2 GHz
- RF in at 57.64 GHz
- Bias voltage +0.7 V
- Conversion Loss 19 dB

Up-convert mode:

- LO +10 dBm at 56.2 GHz
- IF in at 0.3 GHz
- Conversion Loss 19 dB
- The diodes were unbiased, or slightly reverse biased for best results.

The LO – RF isolation in both cases was measured at 26 dB which validates the reduced size ratrace design adopted in the design.

5.7.2.2 Packaged Diode Mixer Characterisation

Similar performance to that measured on wafer was obtained for the balanced diode mixer when characterised in a customised package – the package design being very similar to that described earlier for the balanced FET mixer, again with the main distinction being that only a single IF connection is involved in the case of the diode mixer. The only difference of significance was that the measured conversion loss in the up conversion mode was 2 dB lower, at 17 dB, than the corresponding measurement for the same circuit when operated as a down converter. Again, this can be viewed as a validation of the packaging approach. It should be noted that the diode mixer is not susceptible to bias oscillations and hence no significant performance improvement is expected by further enhancing the package.

6. DISCUSSION

There have been a number of key learning points associated with the mixer designs studied during the course of this work.

In terms of circuit topology, a number of architectures have been examined and these have been compared both from a fundamental standpoint and from a simulation perspective. The need for a balanced approach to ensure good circuit isolations and also to facilitate the suppression of some of the unwanted mixing products has been described. In this context, the LO and RF being applied to the device's gate became the natural topology for fabrication.

In terms of the disappointing conversion loss performance realised with the balanced mixers in comparison with the simulated values, some comments are appropriate. Firstly, consider the diode mixer that exhibited conversion losses in both down and up conversion modes considerably poorer than predicted by the circuit simulator (7 dB worse for the down converter, 6 dB worse for the up converter). The Schottky diode was implemented in the circuit simulator by short circuiting the source and drain terminals of the H40 pHEMT device model and thereby converting the device into a two-port component. This pHEMT model was extracted, by GMMT, near a pinchoff bias point. Such a bias arrangement may be suitable for some FET mixers, but is by no means suitable for a diode mixer application. Given, on the one hand, that the FET model is generally optimised for negative bias voltages on the gate with respect to the source and drain, and that the current characteristic of primary concern in a FET model is I_{DS}, and on the other hand that in a Schottky diode mixer the anode is typically forward biased and the current of interest is the forward conduction current through the anode (gate), it is reasonable to conclude that the FET model is unlikely to be particularly suitable for use in predicting performance when wired up as a diode. Additionally, given the differing bias regimes, the optimum material layer structures that are appropriate for FET and diode operation are not the same. Finally, the pHEMT model was characterised to 40 GHz and expecting this to extrapolate well to 60 GHz may be overly optimistic. In light of this, it is not very surprising that the measured performance does not agree well with the predicted values. In fact, it is somewhat encouraging that the measured conversion losses show a similar trend to those simulated in the sense that the up converter conversion loss is slightly better than the same parameter for the down converter. It is clear that a specific and accurate Schottky diode model is desirable for a mixer simulation to yield accurate results. Moreover, for improved diode mixer performance to be realised it is likely that an optimised material layer structure would be necessary, as has been developed for this very purpose by UMS. On this basis, the combination of high performance pHEMT and diode devices on the same wafer is a challenge which, if overcome, would be a highly desirable development.

In the context of the balanced pHEMT mixer, the measured conversion loss is approximately 3 dB poorer than predicted. The likelihood is that a number of factors are contributing to this discrepancy. Firstly, as mentioned above, the validity of the pHEMT model is questionable, particularly near 60 GHz, given that the model is being used far above the frequency range over which it was extracted. Secondly, the model is extracted at a specific bias point, and its accuracy over a broader bias range is suspect. Given that the large signal LO has the effect of changing the instantaneous bias point to a considerable extent, a large signal device model that is accurate over a broad range of bias points is desirable. It is possible that the packaging may be contributing some additional loss, in particular the high frequency ohmic losses on the quartz based microstrip transmission lines and non-optimum bond-wire interconnects. Nevertheless, the performance shows promise, and it is also worthy of note that the measured pHEMT mixer conversion loss is about 9 dB better than its diode equivalent (down converter) – the corresponding difference is about 6 dB for the predicted data.

Another learning point of importance is that, for flexibility, a diode mixer is preferable to a FET mixer because it can perform similarly as both a down and an up converter. This dual functionality can be advantageous in situations where time to market is key, or where development resources are limited. The biasing requirements of the diode based design are also simpler. On the other hand, the FET type approach does yield superior performance, in terms of conversion loss, for a given application. However, the biasing input requirements are more complex. The packaging of a FET based balanced mixer is additionally complicated by the need for an IF hybrid off-chip. Not alone does this add cost to the final product but it results in a less attractive solution in terms of size, elegance, and of course introduces an additional potential source of manufacturing error and thereby could have significant yield implications.

The reduced size ratrace is an interesting concept. For certain applications, where the full bandwidth offered by the conventional ratrace structure is not required, this alternative implementation yields similar performance over a restricted band. In fact, as has been shown in this work, the reduced size circuit can in fact yield superior performance due to the reduced ohmic losses associated with the smaller structure. The mixer isolations realised with the fabricated MMICs support the view that this ratrace is an effective solution where minimised circuit size is a key concern.

The original and intuitive FET mixer analytical approach developed in this work comprises a useful aid to help the circuit designer gain an improved insight into FET mixer operation, in particular in the context of identifying the optimum bias point for a given topology. This in turn ensures that the circuit simulator is not merely used as a 'blind' tool, but is instead used by the designer to refine a design that has already been optimised conceptually. Not alone will this lead to more robust designs, but also faster development cycle times and reduced time to market.

7. CONCLUSION

Both FET and diode mixers have been studied, and balanced versions of both have been fabricated and evaluated. The performance achieved in the laboratory falls short of that simulated, and possible reasons for this have been postulated. These include model validity, packaging contributions to loss, and, in the case of the diode design, the suitability of the wafer material layer structure. Nevertheless, the trends identified in simulation are mirrored in measurement, with the balanced pHEMT down converter performing several dB better than its diode equivalent in terms of conversion loss.

It is clear that the H40 process has some capability near 60 GHz, but it is also clear that it is in this frequency range that the process is beginning to reach its limits. Given this scenario, it is not surprising that the model validity is questionable, particularly considering that the models are extracted from measurements to only 40 GHz.

The intuitive FET mixer analysis presented in this chapter is a very useful circuit design aid and provides a convenient way of comparing and contrasting topology approaches at a very fundamental level.

8. LIKELY FUTURE TRENDS

It is clear from an integration standpoint that the diode mixer has significant advantages over its FET counterpart due to the latter's IF hybrid requirement. However, for this to be fully exploited in terms of diode mixer integration in multi-function transceiver type circuits, a prerequisite is that the wafer material be capable of supporting a high performance diode. More exotic material structures will be required to meet this demand, while at the same time supporting the need for the high performance transistor devices required for the low noise and power amplifiers. It is likely that GaAs based wafer fabrication processes targetting the mm-wave frequency range will evolve in this direction over the medium term.

In terms of mixer circuit schematic developments, future mm-wave mixers are likely to be sub-harmonic mixers driven by lower frequency sources, or alternatively conventional mixers with frequency multipliers used in the LO chain. The choice will depend to an extent on the technology used. For instance, FET devices are not usually considered appropriate for oscillator design due to their relatively poor phase noise characteristics. However, HBT devices show real promise in that regard, and a process capable of supporting high performance HBT devices in conjunction with possibly pHEMT or diode devices for mixing would be a very attractive development in terms of the sub-harmonic mixer approach. For the development of transceivers using frequency multipliers in the LO chain, the benefits of a process which can support high performance HEMT and diode devices has already been alluded to. However, the natural extension of this approach is the integration of the oscillator, and the HBT device is a natural choice if supported by the process. In a sense then, the natural progression of mm-wave processes will be towards more integration by the addition of alternative high performance device structures on a given wafer material. The most attractive processes will be those capable of sustaining different device types with high performance thereby facilitating highly integrated functionality, as distinct from a process capable of sustaining a single device with state of the art performance.

Chapter 7

FET FREQUENCY MULTIPLIERS

1. INTRODUCTION

A key constituent in any radio transceiver is a local oscillator source to drive the up and down converting mixers. The nature of this source, in particular in terms of its phase noise characteristics, determines to a great extent the transmission and reception qualities of a system. Silicon based oscillators are generally capable of achieving phase noise characteristics superior to GaAs based designs. This is to a degree historical, as most GaAs based transistors are FETs and the phase noise characteristics of FET based oscillators are inherently poorer than bipolar designs (the $1/f$ noise in a FET is dominated by surface effects; in a bipolar device $1/f$ noise is lower because surface states no longer contribute significantly to the current flow.). As a result, oscillators based on silicon are commonly used in RF and microwave systems. Due to the limited high frequency capability of silicon devices, a common approach for higher frequency systems is to cascade a silicon oscillator with a frequency multiplier. The phase noise characteristics are essentially derived from the fundamental source and are typically far better than those that can be achieved with a fundamental GaAs design. At higher frequencies, the development of a fundamental GaAs source can become difficult or even impossible to achieve and, in these circumstances, a frequency multiplier becomes inevitable if system designs incorporating such solutions as Gunn or IMPATT devices are to be avoided.

For many years, a varactor device has been commonly used for frequency multiplier applications. At high frequencies, Schottky varactors are usually employed. When suitably designed and fabricated, these devices can have very high cutoff frequencies. This feature makes them ideally suited to mm-wave (and beyond) applications. Typically, such varactor-based multipliers are implemented in waveguide technology.

A frequency multiplier solution based on a FET type device would be a very attractive alternative as it offers the potential for being integrated with other elements of a monolithic transceiver, including (potentially) the mixers, and the receive and transmit amplifiers. A varactor multiplier can be realised with excellent harmonic conversion efficiency. Achieving similar efficiencies in a FET based design is a challenge. Moreover, a FET based design requires more of an insight into the conversion mechanisms in order that good conversion efficiencies can be realised.

In this chapter, various aspects of FET multiplier development are discussed. Firstly, an original analytical approach for single-ended FET multipliers is presented in Section 2. A simple, but effective, piecewise linear model is assumed for the generic FET in this section. A single unified analysis is developed which predicts the harmonic generation properties of this generic FET for arbitrary bias and drive conditions. The predictions of this formulation are presented in the form of contour plots. Conditions for optimum frequency doubler and tripler performance are considered, and potentially interesting operating points for multipliers are discussed.

Specific mm-wave MMIC multiplier designs developed during the course of this work will be presented in Section 3 of this chapter. These designs have been developed to address the requirements of two distinct mm-wave transceiver markets: firstly, the 57 GHz short-hop unregulated radio link market as introduced in Chapter 4; secondly, the broadband interactive MVDS market in the 40 GHz frequency range. The circuits involved are a frequency tripler to 56 GHz for the former requirement, and a frequency doubler to 40 GHz for the MVDS application. Both of these MMICs have been designed on the GMMT H40 pHEMT process. The performance realised with the 56 GHz tripler is compared both quantitatively with simulations and qualitatively with the expected trends as predicted by the analysis in Section 2 of this chapter. A different approach is adopted in the case of the 40 GHz doubler, with a buffer amplifier being integrated at the output of the doubler stage. The evolution of this doubler design from a raw pHEMT device to the final fabricated circuit is described in detail. These two frequency-multiplying approaches are compared and contrasted. In general terms, the doubler performance is superior to that achieved with the tripler.

An inevitable undesirable output of a frequency multiplier is unwanted harmonics. Filtering is commonly employed to reject these tones. However, combining more than one multiplier using suitable hybrid circuits can improve the unwanted harmonic rejection properties of a frequency multiplier. Such balanced multipliers also allow increased output power levels to be realised, and this can be particularly important at high frequencies where the power output capability of a single device is constrained. Balanced multiplier considerations are discussed in Section 4 of this chapter. In particular, a novel circuit implementation for a flexible balanced harmonic generator is proposed in this section. The proposed concept involves a balanced multiplier and an electronic phase shifter, and by suitably adjusting the phase shift and the individual multiplier bias arrangements (all these adjustments being amenable to electronic control), a multiplier of any order is theoretically feasible.

Finally, Section 5 of this chapter consists of a study of the "art of the possible" in the context of the likely performance achievable from a 100 GHz multiplier chain given a state of the art high frequency MMIC process. This represents what the authors believe will be a key trend in high frequency MMIC developments. The realisation of significant power at high frequencies from monolithic multiplier chains will open up a host of new volume applications for MMIC technology; the automotive collision avoidance system developments at 77 GHz would be of particular interest in this regard.

2. THEORETICAL ANALYSIS OF SINGLE-ENDED FET MULTIPLIER

2.1 General Background

A large volume of work has been published on the topic of FET type frequency multipliers. Theoretical[65, 66] and experimental[67-70] studies have been presented. The general inference from theoretical work, that FET devices are primarily suited to doubler and (to a lesser extent) tripler applications, has been largely borne out by experimental findings. The complexity of the expressions used to model the drain current dependence in the theoretical analyses has varied significantly: Gopinath et al[65] utilised a detailed I_{ds} model which facilitated a study of the influence of various aspects of the non-linear device on its multiplying behaviour; Maas[66], on the other hand, applied a generic piecewise linear transconductance (g_m) model (PLT), where only I_{ds} clipping effects are considered, to develop generalized multiplier biasing and drive criteria. Maas' approach is consistent with the findings of Gopinath et al[65] that the I_{ds} clipping contribution is the most significant factor affecting harmonic generation.

However, recent developments[71] have demonstrated an alternative tripler bias arrangement which has the potential to perform significantly better than would be predicted by the earlier theoretical models. This improvement has been explained analytically (using the PLT model), but it has not been shown conclusively whether it is the true optimum or whether yet further improvements are possible. Moreover, it also casts some doubt on the conclusions drawn in the conventional analysis regarding frequency doublers.

In this section, we examine a combination of the best of the Gopinath et al and the Maas techniques. The goal has been to study, and if necessary refine, the multiplier design criteria of Maas, by applying the PLT model in a more generalized and more extensive harmonic analysis. The criteria for third harmonic optimisation will be compared with the findings of Fudem et al[71]. The application of the PLT model facilitates a good fundamental understanding of the key parameters influencing the harmonic conversion capability of a FET device. In order to take real device characteristics into account, the Angelov device model[29] is used to assess the validity of the design criteria derived using the PLT model.

Finally, despite the fact that the primary value of the analysis in this section is the intuitive insight it provides, and not quantitative results, model validation is carried out in Section 3 of this chapter for a fabricated GaAs MMIC frequency tripler with an output at 56 GHz. The measured conversion efficiency versus input power characteristic for this circuit is compared with the predictions of both the PLT and the device specific (Angelov) models, with acceptable agreement being achieved. This 56 GHz tripler is a key component of the 57 GHz transceiver which was studied by the authors and is discussed in detail in Chapter 8.

2.2 Review of Existing Published Analyses

2.2.1 Gopinath et al[65]

This work concentrated on frequency doublers in particular, and derived a number of conversion efficiency values based on a specific I_{ds} model. The strategy used was to study the contribution to second harmonic generation of the main non-linear factors involved, including C_{gs} non-linearity, I_{ds} clipping and the V_{gs} - I_{ds} transfer characteristic. A key finding was that I_{ds} clipping is the primary contributor to second harmonic generation.

Gopinath et al also conclude that the optimum bias point for doubler operation is either near pinchoff, or else near gate forward conduction. In both cases, the resultant clipping of the waveform leads to significant harmonic generation.

2.2.2 Maas[66]

Mass approached this subject in a more fundamental manner by making the assumption that g_m is piecewise linear. The only non-linear effect considered is I_{ds} clipping. This PLT model is not strictly valid for a MESFET (and even less so for a HEMT due to the g_m peaking caused by a neutral region developing in the doped wide-bandgap material as the gate bias is increased), but nevertheless is very useful in providing an insight into the principal sources of harmonic generation and in developing a first order estimate of the optimum drive/biasing combination.

Maas also assumes that only pinchoff-based clipping occurs, and that forward conduction effects are avoided. On this basis, a Fourier analysis is carried out, from which a set of harmonic coefficient curves is derived. In the analysis, Maas assumes that the non-linear device is biased near pinchoff, and that drain current flows when the superposition of the bias and the applied ac signal brings the instantaneous gate voltage above pinchoff. The fraction of the full ac cycle for which current flows is referred to as the *aperture*, and this is determined by the gate bias and the ac drive level. This mode of operation is essentially similar to that of a Class-C type amplifier. The net result is that the drain current flows in distinct pulses, and Maas makes the further assumption that these individual pulses are themselves sinusoidal in shape (this sinusoid having the same period as the applied signal only for an *aperture* of *0.5*). In the authors' new approach to follow, a more exact analysis is presented which shows that in most practical cases, the sinusoidal pulse assumption is justified.

Based on the results of this analysis, it appears that the FET type device is suitable for doubling and tripling requirements. For higher harmonic numbers, the conversion efficiencies appear poor.

2.2.3 Fudem and Niehenke[71]

Fudem and Niehenke have developed balanced active triplers based on a new biasing arrangement. Instead of adopting the Maas bias scheme, which would demand an aperture in the region of *0.25* (see Fig. 7-2), they have instead opted for a bias point midway between pinchoff and the onset of forward conduction, consistent with a Class-A amplifier. Having biased the device thus, it is then over-driven by a large amplitude ac signal that drives the device both into pinchoff and into forward conduction every cycle. The resulting drain current waveform is severely clipped at both ends, thereby introducing harmonics. When driven hard, the current waveform resembles a square-wave. Among the interesting properties of an ideal square-wave is the fact that it contains odd harmonics only of the fundamental. The Fourier analysis of the severely clipped waveform suggests that the maximum third harmonic content of the drain current, for a given device biased in Class-A mode, exceeds that which can be achieved with the same device when biased near pinchoff according to Maas' design criterion. Fudem and Niehenke have published measurements on functional balanced triplers biased in this way, thereby demonstrating the viability of this approach.

2.3 New Generalized FET Multiplier Analysis Approach[72]

The conventional output of a theoretical FET frequency multiplier analysis is a combination of gate bias and fundamental RF input gate levels for optimum generation of a given harmonic of the fundamental. Suitable input and output matching networks are then required to extract the available desired harmonic power, while at the same time rejecting the unwanted harmonics (including the fundamental). To a significant extent, these terminations are independent of the device and are generally strongly dependent on the fundamental frequency involved (very often, they consist of reactive shunt stubs which present short circuits at the unwanted frequencies). These terminations do not contribute to the conversion performance as such, and the frequency conversion *capability* of the device can be assessed on the basis of *ideal* terminations.

2.3.1 Refinement of Maas Analysis

A simple representation of the gate voltage and drain current waveforms associated with a device biased near pinchoff, is outlined in Fig. 7-1. The gate dc bias voltage level is V_b, the threshold voltage is V_t, forward conduction commences at V_{fwd}, and the peak value of the current pulses is I_{peak}.

As outlined earlier, the primary assumptions made by Maas are:

- g_m is piecewise linear.
- The drain current pulses are half-wave rectified sinusoids.

A straightforward analysis can now be carried out to determine the Fourier coefficients, I_n, for the drain current waveform in Fig. 7-1. By arbitrarily setting $t = 0$ at the peak of the current pulse, the waveform is even and only the cosine coefficients (let us represent these by I_n) need to be considered. Assume that the pulse width is t_0 centred on $t = 0$. The period of the waveform is T. Then, we find:

$$I_n = \frac{2}{T}\int_{-t_0/2}^{t_0/2} I_{peak}\cos\left(\frac{\pi t}{t_0}\right)\cos(n\omega_0 t)dt$$

$$= \frac{4}{T}\int_{0}^{t_0/2} I_{peak}\cos\left(\frac{\pi t}{t_0}\right)\cos(n\omega_0 t)dt$$

$$= \frac{2I_{peak}}{T}\int_{0}^{t_0/2}\left[\cos\left(\frac{\pi}{t_0} - n\omega_0\right)t + \cos\left(\frac{\pi}{t_0} + n\omega_0\right)t\right]dt$$

$$= \frac{2I_{peak}}{T}\left[\frac{\sin\left(\frac{\pi}{t_0} - n\omega_0\right)t}{\frac{\pi}{t_0} - n\omega_0} + \frac{\sin\left(\frac{\pi}{t_0} + n\omega_0\right)t}{\frac{\pi}{t_0} + n\omega_0}\right]_0^{t_0/2}$$

$$= \frac{2I_{peak}}{T}\left[\frac{\cos n\omega_0 t_0/2}{\frac{\pi}{t_0} - n\omega_0} + \frac{\cos n\omega_0 t_0/2}{\frac{\pi}{t_0} + n\omega_0}\right]$$

$$= \frac{2I_{peak}}{T}\cos\frac{n\pi t_0}{T}\left[\frac{1}{\pi/t_0}\right]\left[\frac{1}{1-2nt_0/T} + \frac{1}{1+2nt_0/T}\right]$$

$$\therefore I_n = \frac{4I_{peak}}{\pi}\frac{t_0}{T}\cos\left(\frac{n\pi t_0}{T}\right)\frac{1}{1-(2nt_0/T)^2}$$

where t_0/T is the *aperture*. Using this result, Maas has developed a set of curves that trace the conversion efficiency to various harmonics as a function of the aperture value.

The above analysis is now modified, by the elimination of assumption (ii) above. Instead, the current waveform is assumed to be a clipped sinusoid, proportional to that part of the gate voltage waveform that exceeds the threshold level. This is a more consistent application of the PLT characteristic. Assuming the peak gate voltage level is V_m, we deduce

$$V_m \cos\omega t_0/2 = V_t - V_b$$

The cosine Fourier coefficients are now derived as follows. A normalisation of current to voltage is assumed:

$$I_n = \frac{2}{T}\int_{-t_0/2}^{t_0/2}\left[V_m \cos\omega t - (V_t - V_b)\right]\cos n\omega t\, dt$$

$$= \frac{V_m}{T}\int_{-t_0/2}^{t_0/2}\left[\cos(n-1)\omega t + \cos(n+1)\omega t\right]dt - \frac{2(V_t - V_b)}{T}\int_{-t_0/2}^{t_0/2}\cos n\omega t\, dt$$

$$= \frac{V_m}{T}\left[\frac{\sin(n-1)\omega t}{(n-1)\omega} + \frac{\sin(n+1)\omega t}{(n+1)\omega}\right]_{-t_0/2}^{t_0/2} - \left[\frac{2(V_t - V_b)}{n\omega T}\sin n\omega t\right]_{-t_0/2}^{t_0/2}$$

$$= \frac{V_m}{T}\left[\frac{2\sin(n-1)\omega t_0/2}{(n-1)\,\omega} + \frac{2\sin(n+1)\omega t_0/2}{(n+1)\,\omega}\right] - \frac{4(V_t - V_b)}{n\omega T}\sin n\omega t_0/2$$

$$I_n = \frac{I_{peak}}{\pi}\left\{\frac{\sin([n-1]\pi\, t_0/T)}{n-1} + \frac{\sin([n+1]\pi\, t_0/T)}{n+1} - \frac{2}{n}\cos(\pi\, t_0/T)\sin(n\pi\, t_0/T)\right\}$$

for $n \neq 1$, and

$$I_1 = \frac{I_{peak}}{\pi}\left\{\frac{\pi\, t_0}{T} - \frac{\sin(2\pi\, t_0/T)}{2}\right\}$$

A comparison between the conventional Maas analysis, and the results obtained using our modified approach above, is presented in Fig. 7-2. The results are presented as the harmonic current I_n, normalized to the peak current (I_{peak}), versus the aperture. Traces for the second and third harmonics only are presented. The results of the two analytical approaches are very similar, certainly up to an aperture value of about *0.5*, which corresponds to the device being biased at V_t. For aperture values greater than *0.5*, the results begin to diverge significantly. At an aperture value of *1.0*, which corresponds to a drain current with a full sinusoidal waveform, the refined analysis yields zero harmonic current coefficients as expected, whereas the Maas model suggests non-zero values due to its assumption of half-sinusoidal pulses becoming increasingly invalid. In terms of multiplier efficiency, and optimum aperture values for harmonic generation, both analyses yield similar results.

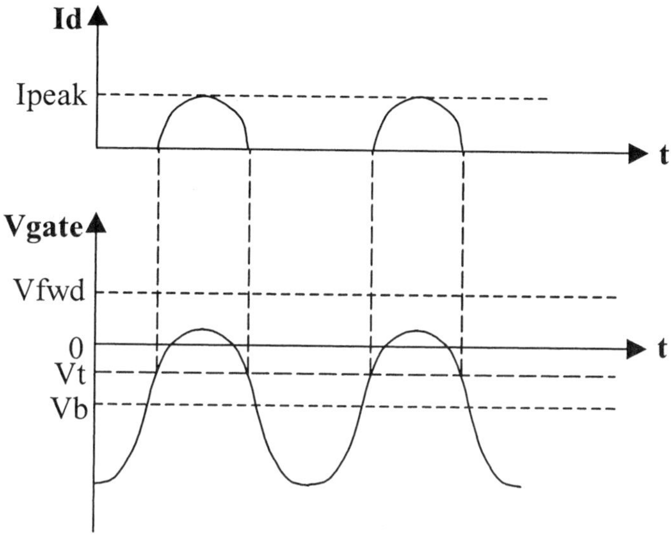

Figure 7-1. Device waveforms for multiplier with pinchoff clipping.

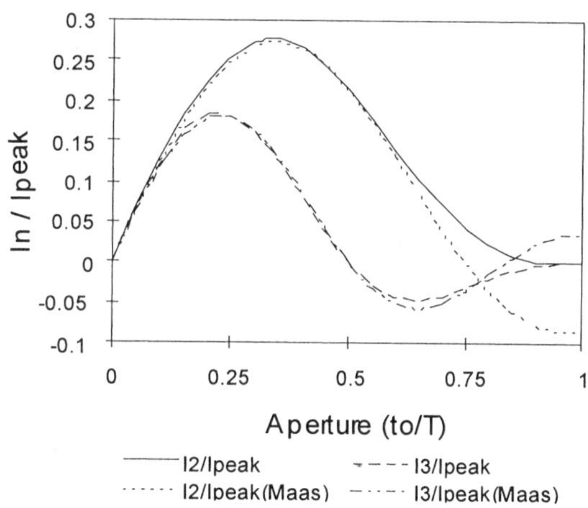

Figure 7-2. Comparison between Proposed and Maas model results.

It can be seen in Fig. 7-2 and it can also be derived from the results above, that the optimum aperture value for a frequency doubler is 0.333, or 120°. The corresponding values for a frequency tripler are 0.25 or 90°.

2.3.2 Justification for a Generalized Theory

As a consequence of Fudem and Niehenke[71], where it has been shown that improved tripler conversion performance is possible by biasing the device differently to the 'optimum' bias suggested by Maas[66], the conclusions of Maas must be reconsidered. Maas' primary concern in opting for the pinchoff region for multiplier biasing was the avoidance of gate forward conduction, as this would introduce significant reliability concerns. Fudem and Niehenke, however, have demonstrated successful tripler operation with circuits in which the clipping effects of forward conduction are exploited.

The analysis proposed by Fudem and Niehenke considers a device biased for Class-A type operation. Consider a situation where the applied ac voltage becomes very large: this causes the current waveform to be severely clipped and an approximately square waveform results. The Fourier terms associated with a perfect square-wave are given by

$$\begin{cases} I_n = \dfrac{2I_{peak}(-1)^n}{n\pi} & ,n \text{ odd} \\ I_n = 0 & ,n \text{ even} \end{cases}$$

Fudem and Niehenke utilized this odd harmonic characteristic of a square-wave in their tripler designs. For a square-wave, the amplitude of I_3 is approximately $0.212I_{peak}$. The corresponding value from the pinchoff clipping analysis (see Fig. 7-2) is approximately $0.183I_{peak}$.

With this in mind, a Fourier analysis needs to be carried out on the generic PLT device. This general analysis should allow for arbitrary bias and drive levels, and facilitate a study of the harmonic generation characteristics of a device operating with pinchoff clipping, forward conduction clipping, or both. Based on the results of this analysis, a contour analysis could be carried out whereby, at a glance, the optimum combinations of bias and drive could be estimated for conversion to a given harmonic.

The theoretical analysis of a multiplier with single-sided clipping has been discussed above. In order to facilitate a generalized study, a Fourier analysis is also required for the situation in which double-sided clipping occurs.

2.3.3 Double-Sided Clipping Theory

Some of the key parameters used in the derivation of this extended model are shown in Fig. 7-3. In order that the results be as generic as possible, a form of normalization is applied. The gate voltage is normalized such that pinchoff effects are observed for values less than zero, and that forward conduction occurs for values greater than *1.0*. In a manner similar to Maas, the Fourier current coefficients are normalized relative to the maximum drain current of the device. This differs subtly from Maas[66], where the normalization was done relative to the peak value of the given current pulse, rather than the maximum value which that peak could possibly attain when driven sufficiently hard. Both normalizations are equivalent when the forward conduction limit is just reached at the peak of the ac cycle.

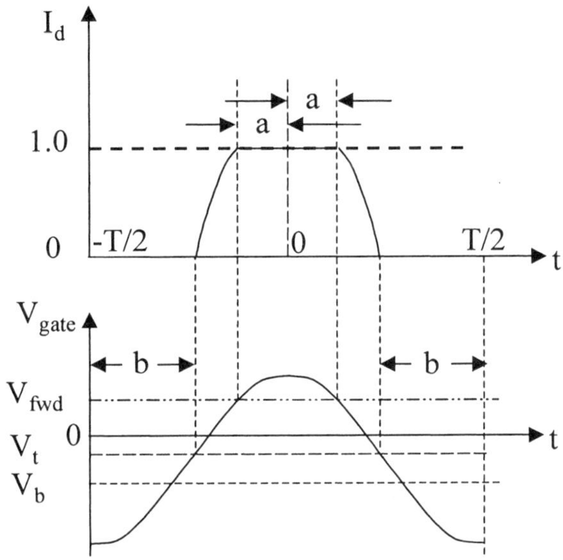

Figure 7-3. Device waveforms for multiplier with double-sided clipping.

In this section, a generalised analysis of the waveforms presented in Fig. 7-3 is introduced. This analysis is arbitrary to the extent that it applies to all situations with double-sided clipping. Subsequently, this result will be consolidated with the earlier single-sided clipping analysis in a coherent generalised theory applicable to cases with double-sided clipping, single-sided clipping and even no clipping at all.

Let the dc and ac gate voltages be V_b and V_{in}, and let the corresponding normalized voltages be V_{DC} and V_{AC}. The normalized values are derived from the actual values, using the expressions

$$V_{DC} = \frac{V_b - V_t}{V_{fwd} - V_t}$$

$$V_{AC} = \frac{V_{in}}{V_{fwd} - V_t}$$

Denote the actual gate bias voltage midway between V_t and V_{fwd} by V_{mid} (= *0.5* normalized).

The clipping durations are given by (see Fig. 7-3)

$$V_{AC}\cos(\omega a) = 1 - V_{DC}$$

$$V_{AC}\cos(\omega b) = V_{DC}$$

For a given V_{DC} and V_{AC}, these equations can be solved for a and b . The normalized current waveform is then given by

$$i(t) = \begin{cases} 0, & -T/2 \le t \le -T/2 + b \\ V_{DC} + V_{AC}\cos(\omega t), & -T/2 + b \le t \le -a \\ 1, & -a \le t \le a \\ V_{DC} + V_{AC}\cos(\omega t), & a \le t \le T/2 - b \\ 0, & T/2 - b \le t \le T/2 \end{cases}$$

A Fourier analysis can then be carried out on this even waveform:

$$\begin{aligned} I_n &= \frac{4}{T}\int_{-T/2+b}^{-a}\left[V_{DC} + V_{AC}\cos\omega t\right]\cos n\omega t\,dt + \frac{4}{T}\int_0^a \cos n\omega t\,dt \\ &= \frac{4}{T}\left[\frac{V_{DC}}{n\omega}\sin n\omega t\right]_{-T/2+b}^{-a} + \frac{2V_{AC}}{T(n-1)\omega}\left[\sin(n-1)\,\omega t\right]_{-T/2+b}^{-a} \\ &\quad + \frac{2V_{AC}}{T(n+1)\omega}\left[\sin(n+1)\,\omega t\right]_{-T/2+b}^{-a} + \frac{4}{n\omega T}\left[\sin n\omega t\right]_0^a \\ &= \frac{4V_{DC}}{2n\pi}\left[-\sin n\omega a + \sin(T/2 - b)n\omega\right] \\ &\quad + \frac{2V_{AC}}{2\pi(n-1)}\left[\sin(n-1)(T/2 - b)\,\omega - \sin(n-1)\,\omega a\right] \\ &\quad + \frac{2V_{AC}}{2\pi(n+1)}\left[\sin(n+1)(T/2 - b)\,\omega - \sin(n+1)\,\omega a\right] \\ &\quad + \frac{4}{2n\pi}\sin n\omega a \quad , n \ne 1 \end{aligned}$$

This reduces to

$$
\begin{aligned}
I_n = \; & \frac{2V_{DC}}{n\pi}\left[(-1)^{n+1}\sin(n\omega b)-\sin(n\omega a)\right] \\
& +\frac{V_{AC}}{(n-1)\pi}\left[(-1)^{n}\sin([n-1]\omega b)-\sin([n-1]\omega a)\right] \\
& +\frac{V_{AC}}{(n+1)\pi}\left[(-1)^{n}\sin([n+1]\omega b)-\sin([n+1]\omega a)\right] \\
& +\frac{2}{n\pi}\sin(n\omega a) \quad , n \neq 1
\end{aligned}
$$

Similarly, it follows that the fundamental component, I_1, is given by

$$
\begin{aligned}
I_1 = \; & \frac{2V_{DC}}{\pi}\left[\sin(\omega b)-\sin(\omega a)\right]+2V_{AC}\left[0.5-\frac{a+b}{T}\right] \\
& +\frac{V_{AC}}{2\pi}\left[-\sin(\omega b)-\sin(\omega a)\right]+\frac{2}{\pi}\sin(\omega a)
\end{aligned}
$$

The DC component can also be derived, but is of no particular interest in this context.

This is a powerful closed-form result making possible the calculation of the Fourier content of a generic FET frequency multiplier current waveform in which double-sided clipping occurs. In the next section, this result will be combined with the earlier single-sided clipping analysis into a consolidated generalised result.

2.3.4 Regimes of Bias/Drive Operation

For a given bias voltage and drive level, a number of distinct regimes of operation can now be identified; these regimes require the application of different analyses (corresponding to either Sections 2.3.1 or 2.3.3 of this chapter) in the calculation of the appropriate Fourier coefficients. This analytical flow is summarized in Fig. 7-4, where the symmetry of the clipping characteristic with respect to the midpoint voltage is exploited. For the case of single-sided clipping, one specific flow through Fig. 7-4 is shown shaded in Fig. 7-5.

The situation summarised in Fig. 7-5 involves a bias point below threshold and an applied ac level with sufficient amplitude to cause drain current to flow and hence lead to clipping at the low voltage excursions (threshold clipping) but insufficient to cause clipping at the high voltage excursions (forward conduction clipping). Other bias and drive combinations can similarly be traced through the flow-chart to identify the appropriate theory to be applied.

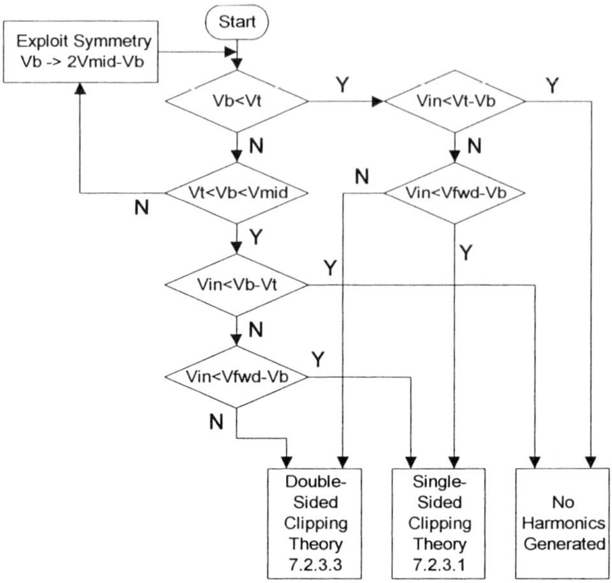

Figure 7-4. Regimes of operation for multiplier analysis.

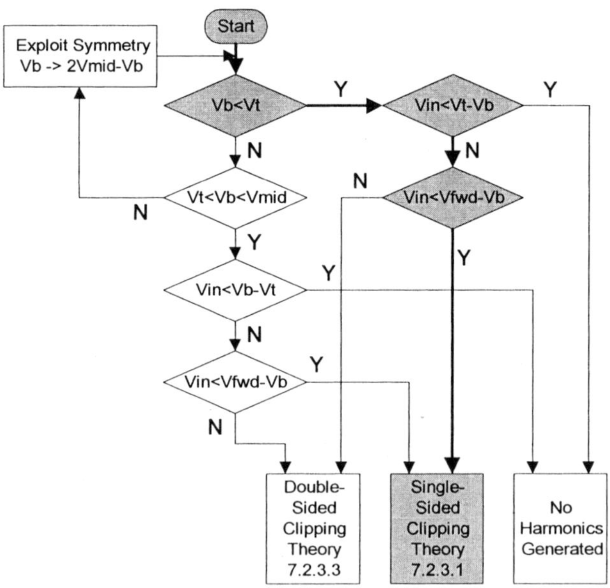

Figure 7-5. Analytical flow for specific (single-sided clipping) bias/drive combination.

2.3.5 Generic Unified Theory Results

The unified theory described above can be used to explore the full range of possibilities, in terms of the normalized bias and drive levels applied to an ideal PLT device, in order to assess the optimum combination for a given multiplier requirement.

The most useful analysis strategy comprises a full sweep of both variables, with the calculation of normalized harmonic current coefficients over a matrix of bias and drive combinations. These results can then be presented and studied as contour plots. These plots facilitate a rapid assessment of the trends in the multiplier characteristics as the input parameters are varied. The multiplier performance for the PLT device is symmetric about a gate bias of V_{mid} and, as a result, the contour plots developed for this generic device only depict the harmonic content for bias levels up to V_{mid}.

Typical contour plots are presented in Fig. 7-6 and in Fig. 7-7. The normalized second harmonic current coefficients are plotted in Fig. 7-6; the normalized third harmonic current coefficients are plotted in Fig. 7-7. Several interesting observations can be made on studying these contour plots.

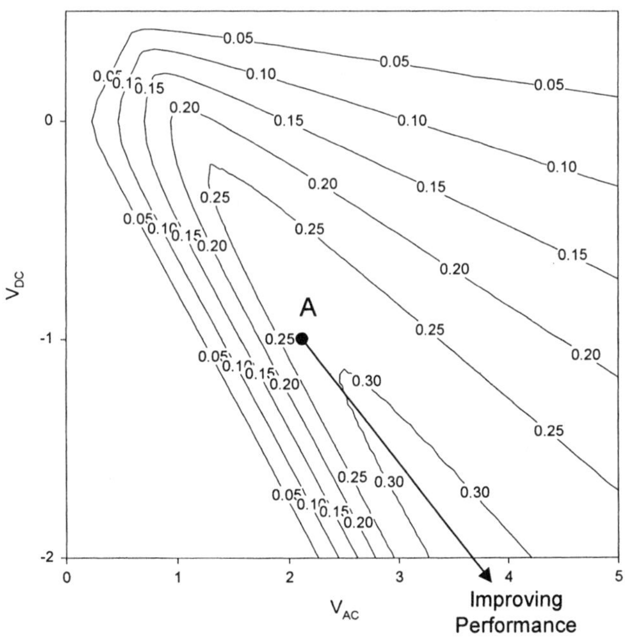

Figure 7-6. Second harmonic contour plot (PLT).

2.3.5.1 Generic Doubler

It can be shown that the optimum condition for second harmonic generation, based on the single-sided clipping analysis of Section 2.3.1 of this chapter, is given in terms of the normalized voltages by $V_{AC} = 2.0$, $V_{DC} = -1.0$, with the corresponding value of I_2 being approximately 0.276. This condition is represented by point A in Fig. 7-6. This results simultaneously in the optimum aperture value and the maximum peak of the drain current pulses, just prior to the onset of forward conduction. It is apparent from Fig. 7-6, and can be calculated using the equations derived earlier, that the value of I_2 continues to increase as V_{AC} is increased beyond *2.0*, with the value of V_{DC} fixed at *-1.0*, and in fact peaks at a value of *0.298*, at a normalized input level, V_{AC}, of *2.3*. This corresponds to an increase in the second harmonic current content of about *8%*. Provided the embedding networks are appropriate, this can also lead to an increase of about *8%* in the second harmonic load power.

It is interesting to study how the optimum values of I_2 and V_{AC} vary as V_{DC} is made more negative. These trends are outlined in Table 7-1. It appears that I_2(max) is heading towards an asymptotic limit. This is confirmed by performing the unified analysis with large V_{DC} and V_{AC} values. It is found that I_2(max) does not increase beyond a certain value, regardless of how large V_{DC} and V_{AC} are allowed to become. This limit can be explained by means of a simple Fourier analysis.

Table 7-1. V_{DC} / V_{AC} trends for maximum I_2 current

V_{DC}	V_{AC} (opt)	I_2 (max)	% Improvement
-1.0	2.0	0.276	-
-1.0	2.3	0.298	8.0
-1.1	2.4	0.300	8.7
-1.2	2.6	0.302	9.4
-1.3	2.7	0.303	9.8
-1.4	2.8	0.305	10.5
-1.5	2.9	0.306	10.9
-1.6	3.1	0.307	11.2
Theoretical Limit		0.318	15.2

It has been mentioned previously that a square-wave contains only odd harmonics. However, by varying the duty cycle of the waveform, so that a rectangular-wave results, even order harmonic content can be introduced. It is easily shown that the second harmonic content of a rectangular-wave peaks when the duty cycle is *0.25*. Under these circumstances, the normalized second harmonic content is given by $I_2(\max) = 1/\pi = 0.318$, which is consistent with the asymptotic limit in Table 7-1. As $|V_{DC}|$ and V_{AC} are allowed to increase, the relative values at the I_2 peak are such that the resulting current waveform is an increasingly close approximation to a rectangular-wave with a duty cycle of *0.25*.

In a practical device, it may be difficult to achieve this level of performance due to breakdown considerations. Nevertheless, it is clear that significantly improved performance (i.e. $I_2(\max) > 0.276$) can be achieved by means of modest changes in the values of V_{DC} and V_{AC}.

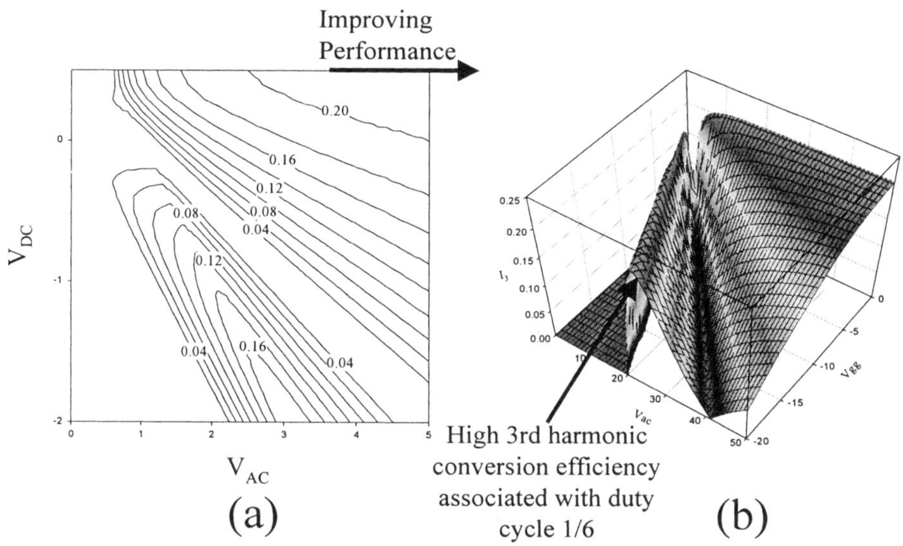

Figure 7-7. Third harmonic contour plots (PLT): (a) normal bias/drive ranges; (b) extended bias/drive ranges.

2.3.5.2 Generic Tripler

In Section 2.3.2 of this chapter, it was explained how over-driving a device biased at V_{mid} leads to a square-wave drain current waveform which has significant third harmonic content. This is consistent with the third harmonic contour plot in Fig. 7-7(a). From the contour plot, it is immediately apparent that by biasing the generic device at V_{mid} (corresponding to V_{DC} = *0.5*), and by applying an adequate ac level, I_3 achieves a value close to the theoretical limit of 0.212, discussed in Section 2.3.2 of this chapter. As outlined by Fudem and Niehenke[71], it is clear that this mode of operation yields superior conversion to the third harmonic than the more conventional pinchoff biasing arrangement.

The study of the Vmid-biased device, driven with a large amplitude ac signal yielding a waveform with an approximately square current waveform and hence a large third harmonic content, prompts a similar analysis of other bias points which would yield rectangular rather than square waveforms in the limit. Consider a rectangular-wave with a duty cycle of $1/6$ which yields a particularly important result. Unlike the square-wave considered previously, this rectangular-wave does contain significant even order harmonics. It is easily shown that the normalized third harmonic coefficient of this waveform is $2/(3\pi) = 0.212$, which is the same at that of the square-wave. In other words, by biasing the generic device such that the drain current waveform approximates a rectangular-wave with a duty cycle of $1/6$, the same optimum value of I_3 can be achieved. In terms of a real device, this would require the device to be biased well below pinchoff, and a high ac drive level would be required. Before configuring a frequency tripler in this way, the breakdown characteristics of the real device in question would have to be considered. A third harmonic contour plot for the generic device, with extended ranges on both the V_{DC} and V_{AC} sweeps, is presented in Fig. 7-7(b).

The previously identified operating point, with V_{mid} biasing, is again apparent. However, another possible operating region, associated with a similar level of third harmonic generation, is evident. This initially manifests itself for a V_{DC} of about *-5.0*, and a V_{AC} of about *7.0*, and it then extends to the right, diagonally down the plot. It is clear that for a given V_{DC}, this high third harmonic conversion is only maintained over a narrow window of V_{AC} levels. Within this window, the relationship between V_{DC} and V_{AC} is such that the resulting approximately rectangular waveform has a duty cycle of about $1/6$. When V_{AC} moves outside this window, the duty cycle deviates significantly from the optimum value and the third harmonic conversion efficiency falls.

The possibility of a $1/6$ duty cycle tripler design point is potentially interesting for a number of reasons. Firstly, the dc current content is significantly reduced, and this reduces the dissipated power. The fundamental content of the rectangular-wave is less than that of the square-wave, and thus the fundamental rejection characteristics are improved. On the other hand, even order harmonics (second harmonic in particular) are introduced; this leads to the need to reject the second harmonic at the output, a problem which does not exist in the square-wave case. Finally, the operating point with the $1/6$ duty cycle requires a large drive power to achieve the necessary V_{AC} levels. The square-wave approach is not as demanding in this regard. However, this is not necessarily a major concern, as low frequency power is relatively cheap and easy to produce. The greatest concern in operating a multiplier in this way is likely to be related to breakdown.

2.3.6 Multiplier Analysis using Angelov HEMT Model

Thus far, the contour plot method has been used as a tool to assess the harmonic generation characteristics of a generic FET device as a function of both bias and applied voltage level. On the basis of the results obtained, some generalized comments have been made concerning the suggested optimum circuit design criteria for a doubler and a tripler.

In this section, a published and commonly used non-linear HEMT model[29] is considered. In order that the analysis be as consistent as possible with that outlined in Section 2.3.5 of this chapter and also in[72], a harmonic balance simulator has not been used in this assessment. Moreover, a harmonic balance simulation could introduce source and load mismatch effects which, though vitally important to actual multiplier performance, would make it difficult to extract specific information on the frequency conversion *capability* of a device as a function of bias and drive conditions. Such capability information is readily available from a Fourier analysis of the device time-domain current waveform derived by applying bias and drive stimuli to the non-linear model. Due to the complex nature of the I-V characteristic of this model, a closed-form Fourier analysis is not possible, and a numerical discrete Fourier analysis was instead carried out to calculate the harmonic content of the drain current. Other aspects of the Angelov model, such as the capacitance elements, were not considered; these of course have an influence on the harmonic generation capability of a given device, but their effects are secondary to the effects of the I-V nonlinearities, as shown in Gopinath[65], and further justified for low frequency considerations by Yhland[73]. As in Section 2.3.5 of this chapter, this Fourier analysis has been performed across a matrix of bias and drive levels. The difference in this case is that the various voltage and current values are not normalized, as a specific device is under consideration; the actual device considered is a Westinghouse 0.5 mm x 0.25 μm power HEMT, with the parameters outlined in Table 5, Appendix 3 of Angelov[74]. The applied frequency has been arbitrarily set to 10 GHz in the analysis; the actual frequency is not significant, as the capacitance effects have been neglected. Having carried out the analysis, the results are again presented in the form of contour plots.

The resulting contour plots for the second and third harmonic coefficients of the output drain current are shown in Fig. 7-8 and in Fig. 7-9. These 'real-device' contour plots are similar in form to those derived in Section 2.3.5 of this chapter, for the PLT generic FET device. For both the frequency doubler and tripler, the same regions of optimum performance are apparent. This demonstrates the validity of the generic unified theory approach.

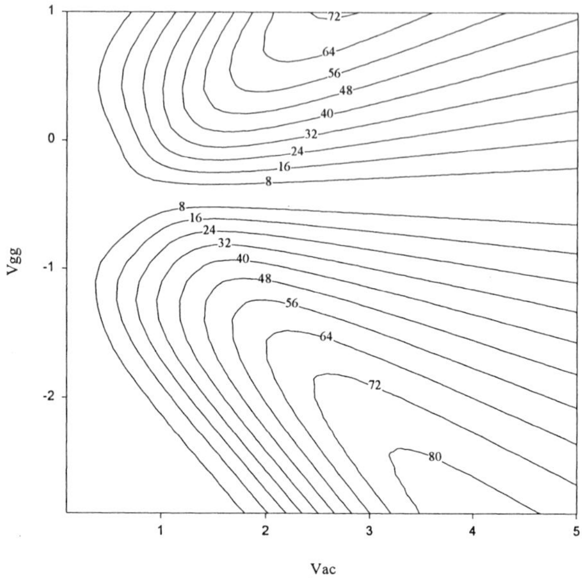

Figure 7-8. Second harmonic contour plot – Angelov device model.

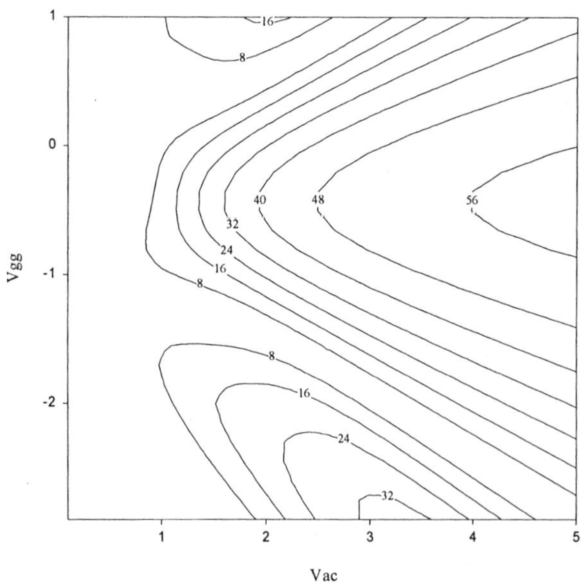

Figure 7-9. Third harmonic contour plot – Angelov device model.

2.3.7 Generic FET Multiplier Analysis Conclusions

The analyses described in Sections 2.3.5 and 2.3.6 of this chapter validate the unified generic theory and demonstrate its value as a simple and effective means of intuitively assessing the harmonic generation capability of a FET. As a closed form analytical tool, the unified generic theory can be used to perform rapid contour plot calculations at a large number of matrix points with fine resolution. Being generic, the analysis is generally applicable to all FET type devices provided the appropriate normalizations are carried out. The 'real-device' approach on the other hand requires a numerical solution, including a complex discrete Fourier analysis. The greater complexity of this approach makes it more expensive to implement; moreover its results are device specific and are not directly 'scalable' to other devices with different characteristics. Of course, for a given device, the 'real-device' approach is likely to yield more accurate results; however for the design of multipliers using modern sophisticated harmonic balance simulation tools, the approximate optimum design suggested by the generic unified analysis is an ideal starting point which the harmonic balance optimizer can refine as necessary to achieve the true optimum design solution. Similarly, the harmonic balance derived optimum design can be checked for consistency with the generic analysis, a useful way of improving the level of confidence in the harmonic balance result.

In the next section, the design of MMIC frequency multipliers is presented. In particular, the measured conversion loss dependence on the applied ac level for a fabricated tripler will be compared with the corresponding predictions of the unified generic and device-specific (Angelov based) approaches. The level of agreement achieved is reasonable, in particular since only the current clipping effects have been considered in these analyses, and the comparison certainly suggests that both approaches constitute valid frequency multiplier design and assessment tools.

3. PRACTICAL SINGLE-ENDED MM-WAVE MMIC FREQUENCY MULTIPLIERS

3.1 Practical Single-Ended MMIC Multiplier Introduction

In Section 2 of this chapter, a fundamental theory of FET frequency multipliers was introduced. In this section, two MMIC frequency multipliers, actually developed during the course of this work, are presented.

The first multiplier is a frequency tripler to 56 GHz. This component is a key element of the 57 GHz transceiver MMIC that lies at the core of this work.

Secondly, the design of a doubler MMIC with an output at 40 GHz will be detailed. This circuit could be of significant interest for the emerging MVDS market in Europe, and elsewhere. The evolution of this doubler design from concept to hardware is presented.

Both designs were fabricated on the GMMT H40 pHEMT process[3]. In common with the other MMIC chips covered in this work, these designs were developed using HP-Eesof's Libra CAE tool[27] on a Unix workstation. Both designs were similar in the sense that a target bias point near pinchoff was selected. The doubler circuit also incorporated an output power gain stage in order to increase the output power capability. One of the MMICs was tested on wafer, and both designs were packaged and subsequently tested at Farran Technology Ltd. The actual biasing point used during the evaluations was adjustable to facilitate the identification of the optimum bias arrangement.

The multipliers were designed for 50 Ω impedances on both the input and output ports. Bias decoupling was included on die but, as is common with such high frequency MMICs, good external decoupling close to the GaAs die is also necessary in the package.

These practical designs comprise useful vehicles to evaluate the validity of the single-ended multiplier theory developed in Section 2 of this chapter. The frequency tripler design in particular is analysed in this context.

Unlike Section 2, where an analysis of a FET doubler was considered before that of a tripler, in this section the practical tripler design will be introduced first. The reasons for this are two-fold. Firstly, this represents the chronological order in which the circuits were developed, and secondly (and far more importantly), the level of complexity attached to the doubler design is more significant than that of the tripler. In fact, the design and evaluation of the tripler provided a key input into the subsequent development of the doubler circuit.

3.2 56 GHz MMIC Frequency Tripler

3.2.1 Frequency Tripler Design Details

A MMIC frequency tripler was designed to generate power at 56.2 GHz[54]. The tripler consists of a 4x40 μm pHEMT device, biased near pinchoff, embedded between reactive stub networks. These networks were designed to provide good input and output matches, unwanted harmonic rejection, and to facilitate bias injection. The design has been fabricated using the H40 process.

The tripler schematic is presented in Fig. 7-10, and the corresponding layout in Fig. 7-11. The input stub, grounded at high frequencies by a capacitor, provides a match for the fundamental input frequency, while at the same time it facilitates bias injection at the gate of the pHEMT device by providing some decoupling. The bias scheme adopted for the tripler was a self-bias arrangement with resistors between the two source terminals and ground. Such a self-bias configuration tends to bias a FET device towards pinchoff, with the resistor value determining how close the device is to pinchoff. The gate is typically grounded in a self-bias scheme; in that case, the gate – source bias is given by

$$V_{gs} = -I_{ds} R_s$$

The larger the value of R_s, the closer the bias point is to pinchoff. With the gate bias pad brought out to an external bias supply, an option effectively to over-write the self-bias setup is made available.

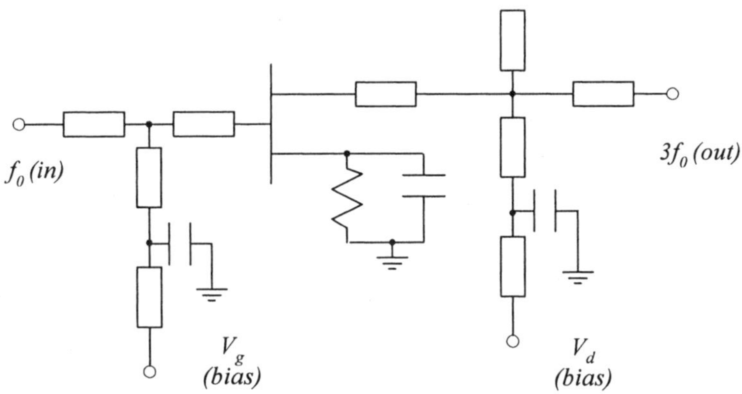

Figure 7-10. Circuit schematic of 56.2 GHz frequency tripler.

On the output side, a double stub arrangement has been employed. These stubs fulfil a number of functions. Firstly, they form an output match for the desired third harmonic signal. Secondly, they implement some filtering of undesirable fundamental and second harmonic leakage to the output. With such stubs, it can be difficult to filter the fundamental without rejecting the desired third harmonic at the same time. This is easily understood if one considers the simplest stub arrangement to reject the fundamental – an open circuit $\lambda/4$ stub meets this requirement. However, such a stub is $3\lambda/4$ long at the third harmonic and tends to reject this frequency as well. As a consequence, the design of the output stub pair involved a compromise in terms of fundamental (and second harmonic) rejection, without excessively loading the third harmonic response. One of the output stubs is essentially a short circuit stub, the high frequency short being provided by a capacitor to ground. This facilitates drain bias injection.

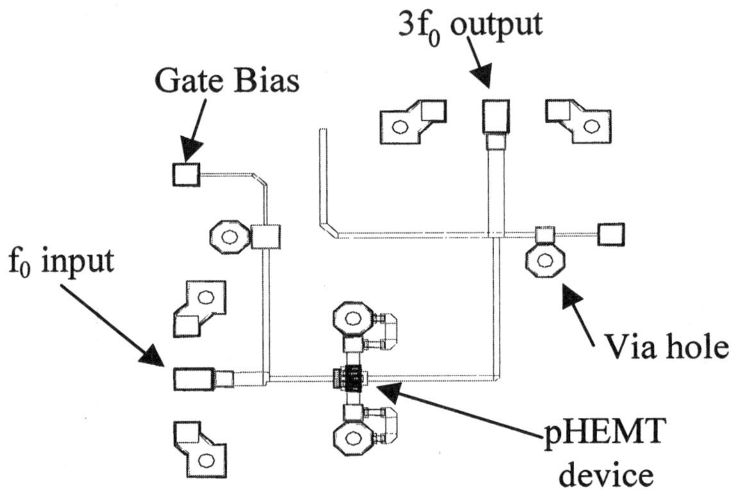

Figure 7-11. Layout of 56.2 GHz frequency tripler MMIC.

The choice of device size is primarily determined by the output power requirement of the multiplier. Higher powers require larger periphery devices to support the higher current demands. When used in a power amplifier type application, for example, the larger devices are necessarily associated with a higher DC current. However, as explained in Section 2 of this chapter, for many FET multipliers the bias point is beyond pinchoff such that the unpumped DC current is practically zero. Of course, when pumped, the instantaneous current flows in pulses as the instantaneous gate voltage swings above pinchoff (see Fig. 7-1), and as a consequence, the average or DC current increases. For many mm-wave MMIC multiplier applications, the DC power requirements are not a major concern. Performance is the key, and the current draw is of secondary importance. This is particularly true in cases where the MMIC approach is being conceived as a possible alternative to, for example, a Gunn diode fundamental oscillator solution – the DC power demands of typical Gunn devices are large (in the range of a few Watts) and monolithic FET based multipliers in conjunction with lower frequency phase locked sources actually become attractive from a power dissipation perspective. Nevertheless, there is no benefit in selecting a device of excessive size, and a device capable of delivering the required output power with some margin is the optimum solution. Devices of excessive size have lower input impedances to the gate, and become more difficult to match to, in particular at mm-wave frequencies where the C_{gs} dominates. This is an interesting distinction between the characteristics of FET devices at relatively low frequencies (for example around the 1 – 2 GHz frequency range) and at mm-wave (and higher) frequencies. At the lower end of the spectrum, the input impedance of a FET gate is high as the reactance associated with C_{gs} is large. However, at higher frequencies, this reactance falls and the input impedance moves towards the short circuit side of the Smith Chart.

The packaging of high frequency circuits must be considered at the design stage, and this is particularly true at mm-wave frequencies, where the parasitics associated with the bond-wire interconnects typically employed can be very significant. It was shown in this work (see Chapter 5) that through careful design, the bond-wire inductance does not become a constraining factor in high frequency circuit design and in fact broadband responses can be achieved. Hence, for the purposes of the MMICs studied in this work, the bond-wire was chosen as the optimum interconnect method, primarily as this bonding process is a mature one and is eminently suited to volume manufacturing techniques. As a consequence, to ensure that the wire parasitics do not severely degrade the circuit performance, the bond-wire inductance was built into the design simulation and optimisation.

For the purposes of circuit design, simulation and optimisation, the H40 Foundry library elements[3], for all components of the design/layout were included in the simulation schematic. Critically, these included the various junction elements including bends, tees and crosses. The inclusion of these elements was important for two reasons. Firstly, the development of the layout from the schematic was more straightforward using the CAE tool's in-built synchronisation feature. Secondly, and more importantly, these elements impact on the expected circuit performance. This in particularly true at high frequencies, where the parasitic effects of these elements are very significant. The development of a mm-wave MMIC circuit design is an iterative process. As the layout is a major component of the design, the schematic development must take the corresponding layout into account at all times. As a consequence, the layout is continually viewed as the schematic evolves to ensure that it is realisable, and that any proximity issues are addressed.

The simulated performance of the frequency tripler is summarised in Table 7-2. The input 18.733 GHz signal is injected at a level of +20 dBm. The predicted output levels for the first four harmonics are shown.

Table 7-2. Simulated harmonic power levels at output of 56.2 GHz frequency tripler

Pin @ 1H 18.733 GHz (dBm)	Pout (1H) 18.733 GHz (dBm)	Pout (2H) 37.466 GHz (dBm)	Pout (3H) 56.2 GHz (dBm)	Pout (4H) 74.933 GHz (dBm)
+20.0	-4.4	-7.8	+8.7	-4.9

The conversion loss to the third harmonic is approximately 11.3dB. The fundamental rejection is 24.4dB.

3.2.2 Frequency Tripler Evaluation

Both on-wafer and packaged testing was carried out on the tripler MMIC. The on-wafer approach is a key requirement to enable the commercialisation of mm-wave MMIC designs at the bare die level. However, the on-wafer testing of circuits at such high frequencies is fraught with difficulties. In particular, as was discussed in detail in Section 2.3 of Chapter 5, the provision of adequate off-chip decoupling can be a major challenge. In this section, an outline of the on-wafer test approach attempted in this work will be presented. The on-wafer measurement test setup is shown in Fig. 7-12.

In this instance, the DUT refers to the wafer-level MMIC multiplier. The input signal is provided by a synthesiser, with a power amplifier being used to boost the power level to the desired +20dBm. On the output side, a power meter was used to measure the third harmonic level. A spectrum analyser was used to determine the frequencies being monitored. The DC inputs were variable so that the optimum bias arrangement could be identified in real-time. The connections from the amplifier and the measurement equipment were made using semi-rigid coaxial cable to ensure good stability of the probes and little likelihood of unwanted movement. The GSG probes used are similar to those described in Chapter 2. A GSG probe with a K-type coaxial terminal was used to inject the fundamental input. A V-type GSG probe was used for the output signal extraction. Needle probes were used to inject the DC control levels to the gate and drain bias pads on the MMIC – the decoupling implemented on these probes is similar to that described in Chapter 5. The layout of the tripler did not include any ground pads adjacent to the DC pads. This precluded the use of a probe card arrangement for the bias injection. A probe card essentially consists of needle probes in a fixed arrangement, customised for a given pad footprint on a die. Typically, it consists of a pair of needle probes in close proximity per DC input to the die. One needle lands on the DC pad, the other lands on an adjacent ground pad. On the probe card, a decoupling capacitor is located between the two needles, as close as is feasible to the probe tips. This capacitor helps minimise the likelihood of any low frequency bias oscillations in the circuit.

This was one key lesson learned in this evaluation, and subsequent MMIC designs were laid out with this probe card compatibility in mind.

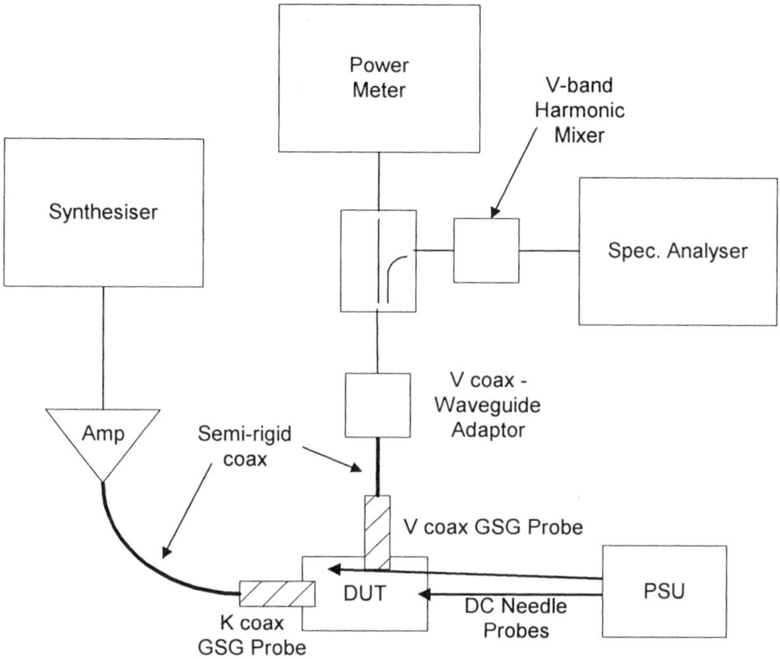

Figure 7-12. 56 GHz frequency tripler on-wafer measurement test setup.

Additional tests were subsequently carried out in a customised package with coaxial connectors. The packaging method is similar to that employed for the LNA and Mixer packaging as described in previous chapters. Essentially, bond-wires were used to interface between the bond pads on the die and external 50 Ω microstrip transmission lines on quartz. Both the GaAs die and the quartz material were epoxied onto the metal base of the package. Such conductive epoxy is suited to relatively low power applications. For dice with higher power dissipation, such as power amplifiers, solder is often used instead of epoxy due to its better thermal characteristics. The feed-throughs from the external coaxial connectors to the internal interconnect to the quartz lines were implemented using appropriate glass beads supplied with the coaxial connectors. In the case of this tripler, a V connector was used for the output signal, whereas a cheaper SMA connector was employed for the lower frequency input signal. Chip capacitors were also epoxied on the metal base, adjacent to the MMIC. Bond-wires were then used as the connections from the DC bond pads on the

MMIC to these capacitors, and then subsequently to bias supplies. These capacitors were necessary to provide additional bias decoupling. Implementing similar value capacitors on die would waste a large amount of expensive GaAs real estate.

The test setup used in the characterisation of this packaged multiplier is shown in Fig. 7-13. It is similar to that shown earlier for the corresponding on-wafer test arrangement. In this case, however, the DUT refers to the packaged multiplier, with coaxial terminals for the input and output signals, and DC filtercons (DC feed-throughs with integrated low-pass filters) for the bias injection.

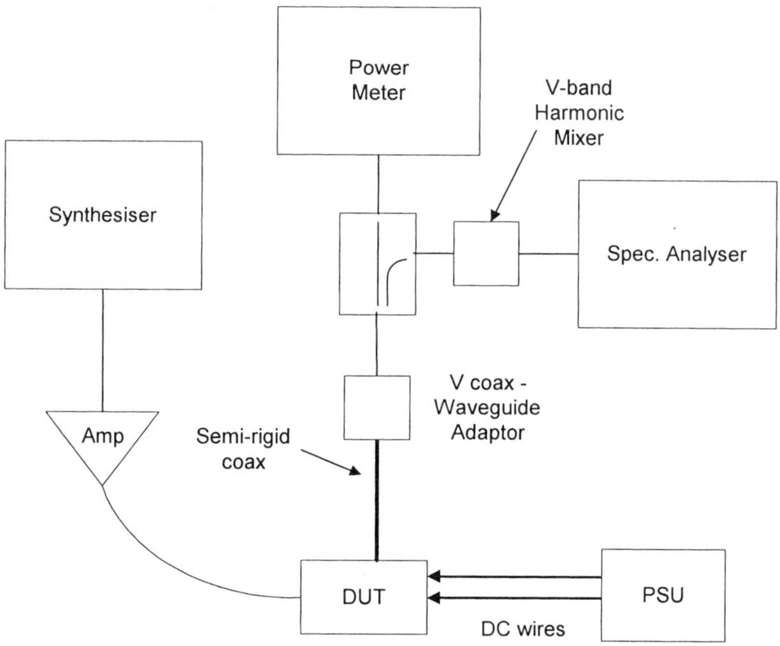

Figure 7-13. Packaged 56 GHz frequency tripler test setup.

Similar results were obtained for the on-wafer and packaged tripler tests. A minimum conversion loss of 14.1 dB was measured (11.3 dB predicted), under optimum bias/drive conditions, for this circuit. The agreement between the on-wafer and packaged data is not unexpected – due to its biasing, the tripler is not susceptible to oscillations and thus the difficulties associated with providing adequate decoupling close to the MMIC for on-wafer testing do not arise in the case of the tripler. Such performance would by no means be considered state of the art. However, at the time of the design, it did represent the highest frequency tripler designed using the GMMT H40 process. The fact that the discrepancy (2.8 dB) between the measured and simulated conversion efficiencies was reasonably small is also encouraging. It should be re-emphasised that the GMMT foundry models were extracted from characterisations carried out to 40 GHz, and the assumption that the models extrapolate well to 60 GHz would have to be considered questionable. It must also be borne in mind that the f_t of the H40 process is in the region of 70 GHz, and circuits designs close to this frequency should be considered a challenge.

The conversion loss has been characterised as a function of the input power, and these results have been compared with the corresponding predictions using the Angelov model (the foundry device's threshold voltage, *-1.4V*, is similar to that assumed for the Westinghouse device for which the Angelov parameters are available[74]). In order to eliminate all other factors, so that the input power dependence of the conversion loss can be compared in isolation, the measured conversion loss values have all been offset by 4 dB so that the resulting optimum conversion loss is consistent with the minimum conversion loss predicted by the Angelov model analysis described in Section 2.3.6 of this chapter. The input power values have also been modified to account for input losses due to the packaging arrangement, and an imperfect input match. Using the threshold voltage value, the results of the generic unified theory analysis (Section 2.3.5 of this chapter) can be *de-normalized* and these also have been compared with the measurements. The resulting comparison is presented in Fig. 7-14.

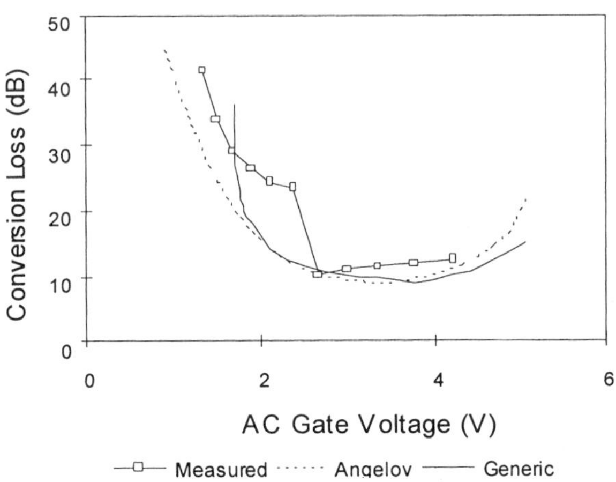

Figure 7-14. Comparison between measured and predicted tripler conversion loss versus input level.

The agreement obtained is reasonable, especially considering that only the current waveform clipping non-linear effects have been considered. Note that the conversion loss associated with the generic model becomes infinite when the ac gate voltage becomes so small that the peak gate voltage is below the threshold. The Angelov model, with its more complex current formulation, yields a softer turn-on in this regard.

It should of course be pointed out that the bias point studied for this tripler circuit is in the pinchoff region. As shown in Section 2 of this chapter, a Class-A type biasing scheme has the potential to yield superior tripling efficiency. However, due to the self-bias arrangement implemented on the MMIC, such Class-A biasing could not be studied in this case.

3.3 40 GHz Frequency Doubler

3.3.1 Frequency Doubler Design Details

A MMIC frequency doubler has been designed to generate power near 40 GHz[75]. The multiplier stage consisted of a 2x60 μm pHEMT device, again biased near pinchoff, embedded between reactive stub networks. These networks were designed to provide good input and output matches, unwanted harmonic rejection, and to facilitate bias injection. This design has also been fabricated using the GMMT H40 foundry process. The pinchoff bias arrangement is entirely consistent with the generic contour plots presented in Section 2 of this chapter. Care was taken to ensure that the limitations in the biasing flexibility of the tripler circuit in Section 3.2 of this chapter, were not repeated in this doubler design.

The evolution of this single-ended multiplier is now described in some detail.

3.3.1.1 Design Evolution

In this section, the evolution of the single-ended 40 GHz frequency doubler is detailed. The evolution is traced from an idealised first-principles schematic, through varying degrees of complexity to a complete single-ended design. Subsequently, two approaches to incorporating this multiplier in a design that is suitable for fabrication are considered, and the most appropriate is selected. The target output frequency range of the doubler is the band about 40 GHz, this being useful for a number of applications, particularly the potentially high-volume MVDS market. For the initial design evolution study, an input power level of +20 dBm was assumed. Subsequently, in the final design simulation and optimisation, this was reduced to +10 dBm to be consistent with the power levels available in emerging MVDS developments.

The evolution is considered in a number of steps. Firstly, the biasing of a device for best doubler performance is considered. This is assessed in the context of the generic theory presented earlier in Section 2 of this chapter. Next, suitable input and output terminations are discussed. These are introduced to reject unwanted harmonics, and at the same time they allow the propagation of the desired signals. The parasitic effects of a laid out form of the design are then taken into account. The next step involves utilising this approach in either a balanced design or in conjunction with an output amplifying buffer stage to enhance the output power capability and improve the unwanted harmonic rejection characteristics. The latter approach in particular facilitates a simplification of the single-ended design. It also results in a smaller area requirement and is selected as the design most suitable for fabrication.

3.3.1.2 Optimum Theoretical Biasing

As discussed in Section 2.3.1 of this chapter, it can be shown that the optimum bias point for a frequency doubler corresponds to an aperture value of 0.333 or 120° - see also Fig. 7-2. As shown in Table 7-1, further enhancements in the second harmonic output can be realised by allowing double-sided clipping to occur. From the analysis in Section 2.3.5.1 of this chapter, what this means in practical terms, is that the device gate needs to be biased at a voltage that is more negative than the pinchoff voltage.

Knowing the pinchoff characteristics associated with the H40 pHEMT devices, a single device (2x60 μm) was biased with an aperture of approximately 120° and a V_{DD} of 3V (V_{GG} = -3.4 V). This gate bias level was calculated assuming a threshold voltage of -1.4V and a gate forward conduction limit of +0.6V. The source was grounded and a negative bias applied to the gate. Ideal chokes were used for bias injection. A 20 GHz signal was applied in simulation to the input of the device (gate) and the output power at both 20 GHz and 40 GHz monitored. No matching was employed in this initial simulation. The results of this simulation, carried out over a range of input power levels, are shown in Fig. 7-15.

A number of interesting points can be made. Firstly, for low input levels, the second harmonic output is rising approximately twice as quickly as the fundamental. Secondly, the fundamental leakage is relatively high, being only a few dB below the applied power at high input levels. Thirdly, the optimum output at 40 GHz is +5 dBm for an incident power level of +18 dBm at 20 GHz.

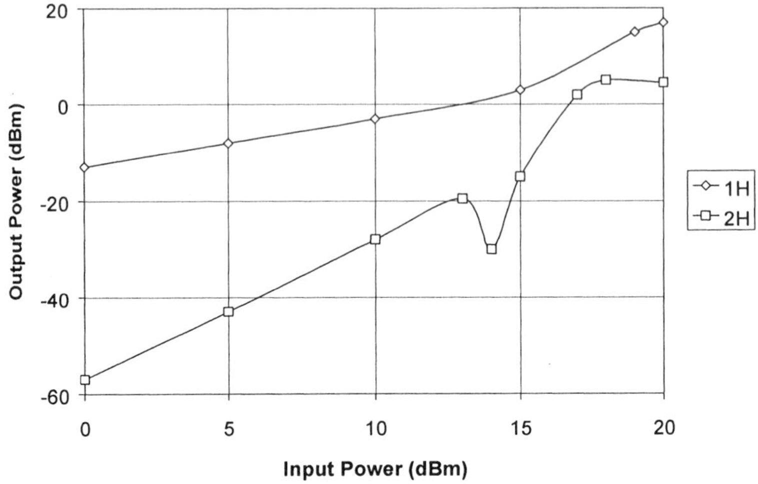

Figure 7-15. Simulated output harmonics (dBm) versus input power (dBm) for a 2x60 μm H40 device biased with an aperture of 120° - no matching employed.

The cause of the dip in the second harmonic trace for an input power level of about +14 dBm is not clear. The most likely cause is that some second harmonic cancellation is occurring because power at this frequency is being generated by both a second order and higher order mixing mechanisms.

3.3.1.3 Input Matching Network

In theory, the input network shown in Fig. 7-16 has very useful characteristics for a doubler. At the fundamental (f_{in}), the reactive effects of the short and open circuit stubs cancel and hence the fundamental is unaffected by their presence. At the second harmonic, the open circuit stub presents a short circuit at node A and thus provides a reactive termination to the gate of the FET at the output frequency. This has the effect of reflecting any output signal at the input port back into the transistor, thereby enhancing the overall output characteristic. Of course, there are also many other combinations of one or two stubs that can yield appropriate terminations at the various harmonics, and the circuit in Fig. 7-16 is only one example. It should also be pointed out that the circuit in Fig. 7-16 is particularly suitable for instances where no additional matching is required at the fundamental (hence the cancellation of the effects of the stubs is key). If fundamental matching is required, then the length of the short circuit stub can be adjusted.

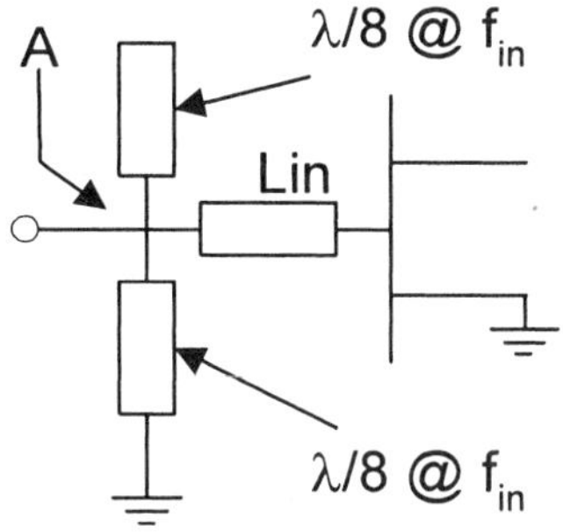

Figure 7-16. Input match circuit for the FET frequency doubler.

The input network in Fig. 7-16 was included in front of the 2x60 μm device considered earlier, and with an applied level of + 20 dBm at 20 GHz, the line length Lin was swept from 0° to 100° of phase length at 20 GHz. The resulting output characteristics are shown in Fig. 7-17.

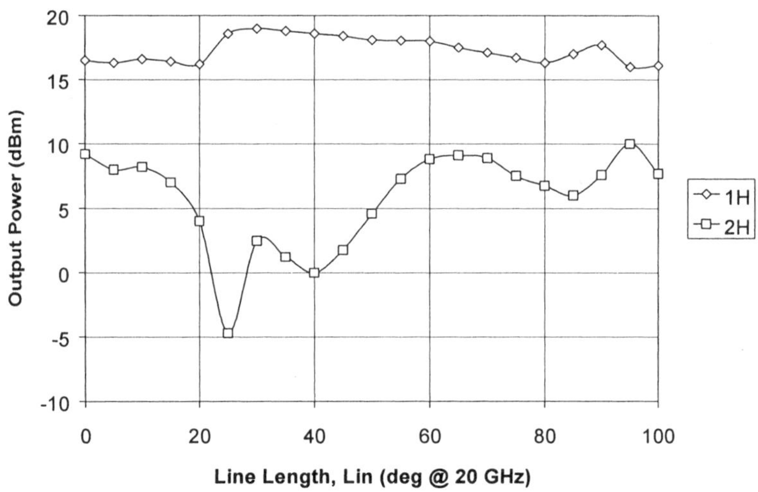

Figure 7-17. Simulated output harmonic levels (dBm) versus Lin for incident +20 dBm @ 20 GHz.

From the traces in Fig. 7-17, it is apparent that an output of about + 10 dBm can be expected at 40 GHz when the input match circuit is introduced, with Lin = 96°. This corresponds to an enhancement of about 5 dB over the earlier result in Fig. 7-15 when no input matching arrangement was involved.

3.3.1.4 Output Matching Network

A similar form of output matching network can be considered, as shown in Fig. 7-18.

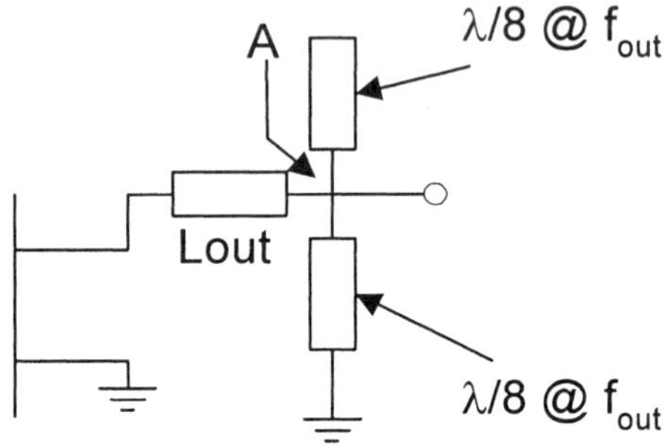

Figure 7-18. Output match circuit for the FET frequency doubler.

As before, the effects of the two stubs cancel, at the output frequency in this case. An additional characteristic of this stub network is that the network tends to improve the match on either side of the band centre. In other words, the bandwidth of the realisable match is enhanced using the two-stub approach.

Again, alternative stub designs can be arrived at – e.g. a simple one is a single open circuit stub whose length is a quarter wavelength at the fundamental.

The output match network (Fig. 7-18) was appended to the circuit simulated in Fig. 7-17. Lin was set to 96° and the output line length, Lout, was swept. Again, the input condition was +20 dBm at 20 GHz. Similarly to before, it was found that further enhancement of the 40 GHz output level can be achieved for a particular Lout value, in this case 130°. An output level of about +12 dBm at 40 GHz is now possible with this combination of device, bias point, input matching and output matching networks.

This latest version of the evolving doubler (with Lout set to 130°), was then simulated versus swept input power at 20 GHz, and the harmonic levels at the output are presented in Fig. 7-19. When the second harmonic output trace in Fig. 7-19 is compared with that in Fig. 7-15 earlier, it is clear that significant enhancements have been introduced.

It should be mentioned here that by simply adding at the output, a quarter wavelength (at the fundamental) open circuit stub, the fundamental leakage to the output can be effectively eliminated without impacting on the 40 GHz level. With this additional stub included, a frequency sweep simulation was then carried out. The input level was fixed at +20 dBm for this sweep. The resulting harmonics at the output are shown in Fig. 7-20.

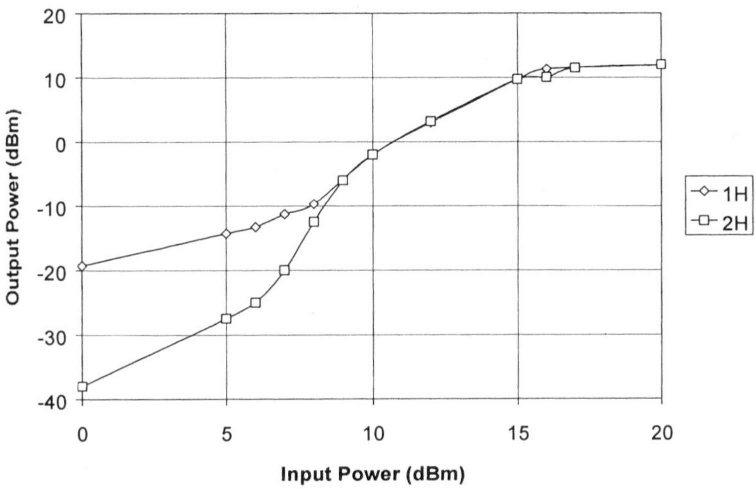

Figure 7-19. Simulated output harmonic levels (dBm) versus input power (dBm) for circuit described in text.

Here, the combination of optimum output at the second harmonic and excellent fundamental rejection at the band centre is evident.

3.3.1.5 Optimised Single-Ended Design

Next, the circuit described thus far was refined. This involved a number of changes, including an output impedance matching quarter-wave transformer, that were introduced for additional performance. The function of this transformer is to provide the optimum load impedance in terms of the maximum voltage and current swings consistent with the device I/V characteristic and supply voltage. Through simulation, it was found that an impedance of about 60 Ω was optimum for the quarter-wave section. The parasitic effects of tees and crosses were also built into the simulation. The key elements of the design were then optimised using the Libra CAE simulation tool[27] and the approximate optimised values are indicated in Fig. 7-21.

The simulated performance for this design is presented in Fig. 7-22. This suggests that an output power of about +10 dBm at 40 GHz can be expected from this circuit when +20 dBm is applied to its input port at 20 GHz.

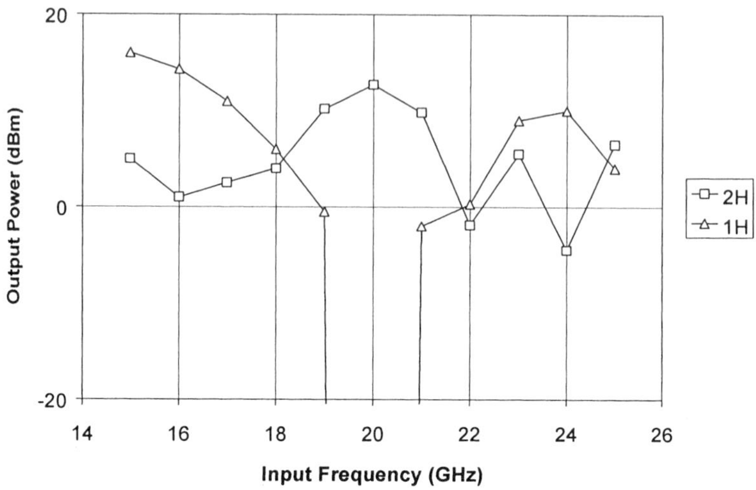

Figure 7-20. Simulated harmonic levels (dBm) versus frequency for circuit with a quarter wavelength (at the fundamental) open circuit stub – Input power +20 dBm.

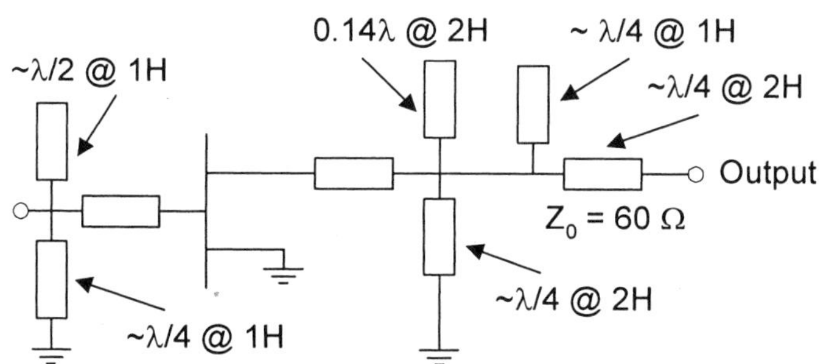

Figure 7-21. Schematic for optimised single-ended 20 to 40 GHz frequency doubler.

3.3.1.6 Practical Realisation of Doubler

The design target for the practical doubler was changed slightly at this stage. The target input power level was reduced from +20 dBm to +10 dBm, to be consistent with the power levels typically available from 20 GHz sources for MVDS applications. The design approach adopted in this instance also differed significantly from that used for the tripler in Section 3.2 of this chapter. In this case, the doubler was implemented as a cascade of a doubler stage (based primarily on the single-ended circuit whose evolution was detailed above) and an output buffer (amplifier) stage. One benefit of this arrangement is that the limited bandwidth of the amplifier necessarily improves the unwanted harmonic rejection characteristics of the doubler stage on its own (and in fact eases the harmonic rejection requirements of the single-ended multiplier), as well as increasing the desired output level. An alternative approach involving a balanced design using two of the single-ended multipliers embedded between a pair of hybrids was also considered, but was quickly abandoned – the area requirement for such a balanced design would be extremely large (involving two of the single-ended doubler circuits plus a hybrid which is typically quite large) and the cost and yield implications associated with this made it an unattractive solution. The buffer amplifier stage was extracted directly from the unit amplifier cell developed for the 30 – 50 GHz LNA described in Chapter 5. The doubler circuit schematic is shown in Fig. 7-23, and the final layout is presented in Fig. 7-24. It can be seen in Fig. 7-24 that the die has been laid out with ground pads adjacent to the DC pads. This is primarily to facilitate the on-wafer characterisation of the circuit. The circuit design/optimisation incorporated the influence of the bond-wires at the input and output. In this case, the output bond-wire is matched back into the driver stage using the broad-banding technique outlined in detail in Chapter 5.

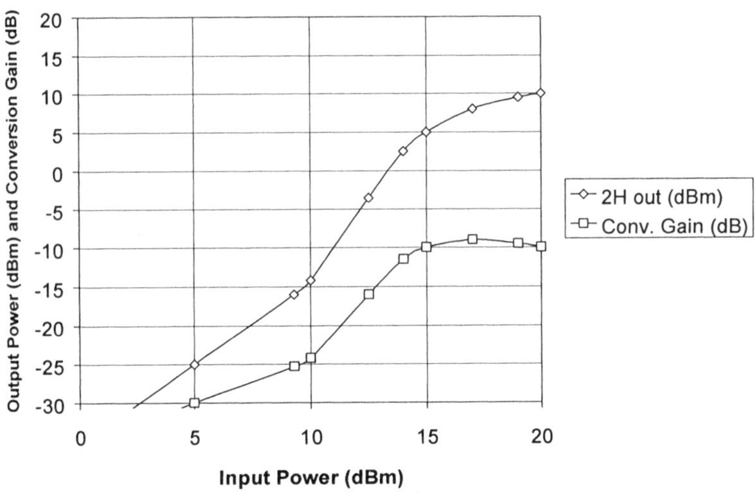

Figure 7-22. simulated second harmonic ouput (dBm) and conversion gain (dB) versus input power (dBm) for the doubler represented in the Fig. 7-21.

The simulated performance of the doubler MMIC is presented in Fig. 7-25. Both the output power at 40 GHz and the associated conversion gain are shown as a function of the incident power level at 20 GHz. The peak simulated output power is +12.6 dBm for an input level of +13 dBm.

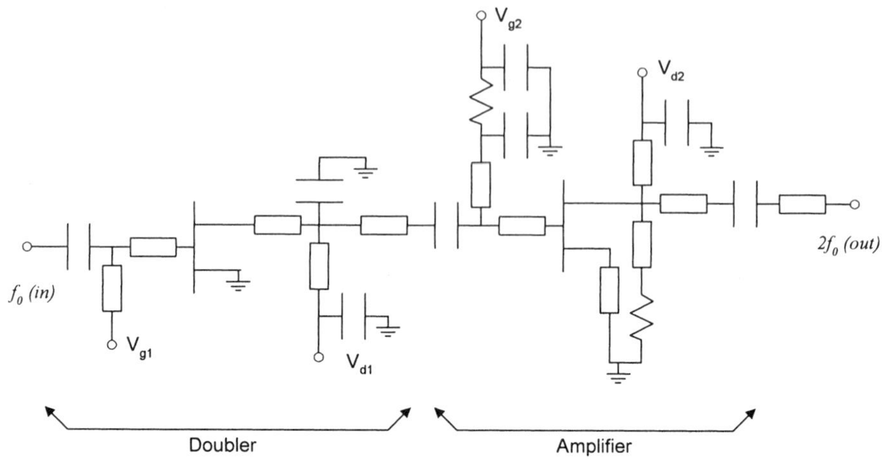

Figure 7-23. Circuit schematic of 40 GHz frequency doubler.

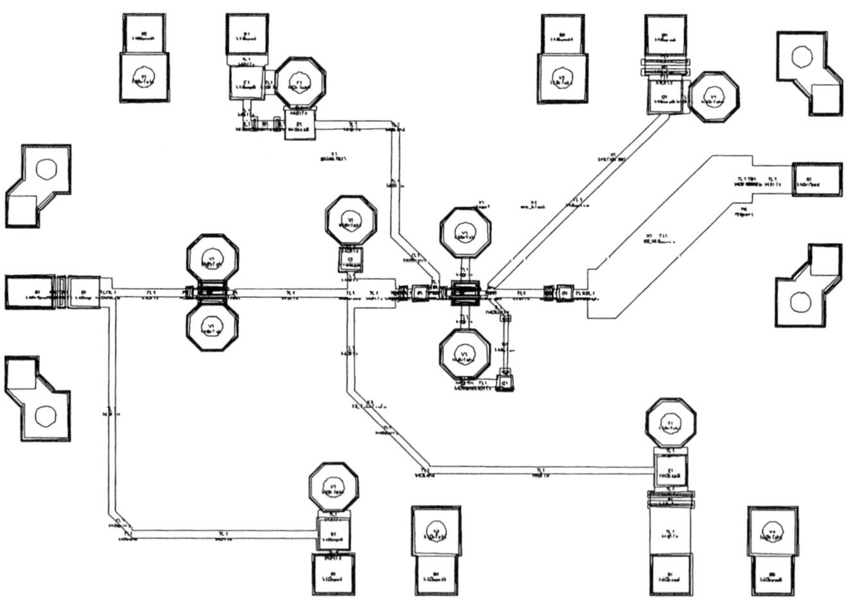

Figure 7-24. Layout of 40 GHz frequency doubler MMIC.

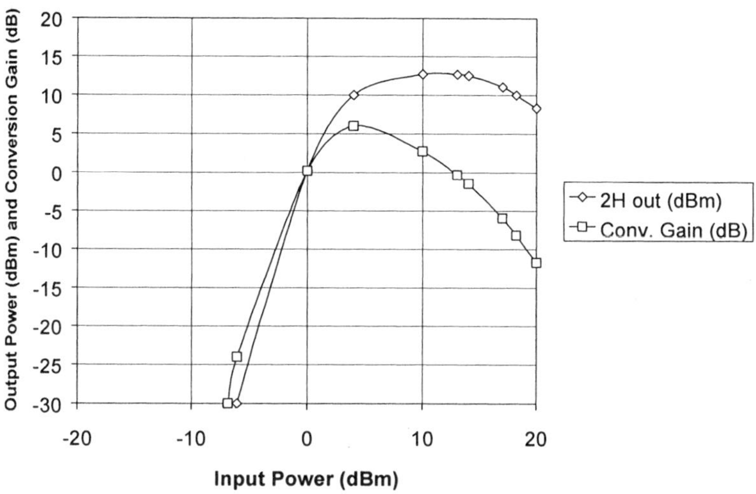

Figure 7-25. Simulated performance of fabricated 40 GHz frequency doubler.

3.3.2 Frequency Doubler Evaluation

The doubler chip was fabricated under the Europractice Multi-Project Wafer scheme on the same wafer as the 30 - 50 GHz LNA described in Chapter 5. As a consequence, on-wafer testing was not an option in this case. The die was characterised in a customised package, a photograph of which is shown in Fig. 7-26. This package is very similar in style to that described in Chapter 5 for the 30 – 50 GHz LNA.

The test setup is represented in Fig. 7-27. The measured performance of the doubler is summarised in Figs. 7-28 – 7-30.

Firstly, a plot of the output power versus output frequency for an applied +9 dBm signal at half the output frequency is shown in Fig. 7-28. Due to test equipment constraints, the input power level was limited to + 9 dBm for this broadband measurement.

A best-case output power of +9 dBm was measured at 40 GHz. It should be borne in mind that this measured performance is realised at the output of the doubler package. It is estimated that the package contributes about 1 dB of output loss, indicating that the true maximum output power from the MMIC at 40 GHz is about +10 dBm. Similarly, the package input loss at 20 GHz is estimated to be about 0.5 dB which means that the true input power level to the MMIC is about +8.5 dBm. This result corresponds to a conversion gain for the MMIC of about 1.5 dB. The corresponding simulated value from Fig. 7-25 is about 12 dBm output power for an input power of +8.5 dBm, a conversion gain of 3.5 dB. This discrepancy of only 2 dB between the expected and simulated conversion performance is highly encouraging.

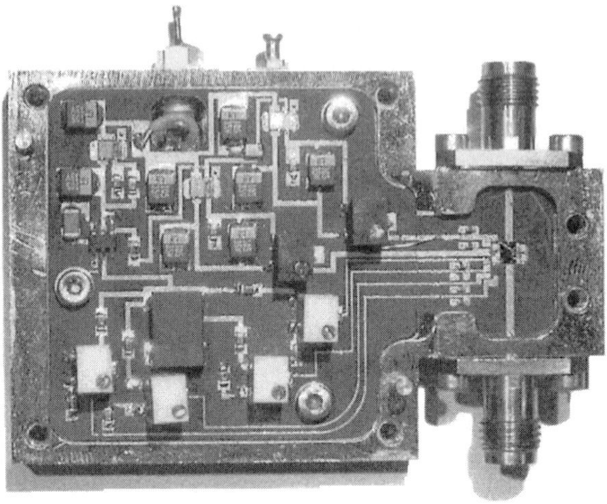

Figure 7-26. Photograph of 40 GHz frequency doubler in customised package.

The overall shape of the measured output power versus output frequency characteristic is quite similar to the simulated response depicted in Fig. 7-20 – the circuit concerned in Fig. 7-20 does not include an output amplifier stage but nevertheless the similarity in the traces is clearly evident. The measured output power is in excess of +5 dBm between 36 – 43 GHz. The corresponding conversion losses are generally in the 0 - 5dB range. This very good performance over a significant bandwidth makes our doubler circuit a good candidate for many applications in the 40 GHz region of the spectrum.

In Fig. 7-29, the measured output power at 40 GHz as a function of the applied level at 20 GHz is shown. The associated conversion gain is also included in this Figure. Again, it must be emphasised that these measured traces are for the packaged doubler. The inferred characteristics for the MMIC involve adding 1.0 dB to the second harmonic power level, and 1.5 dB to the conversion gain result. With this in mind, the maximum inferred output power for the MMIC itself is about +10.5 dBm. This is 2.1 dB less than the maximum simulated level (Fig. 7-25). This difference could be due to a number of possible factors including either greater than anticipated package loss or output buffer compression at lower than expected power levels. The latter in particular is considered unlikely, as the same amplifier stage was found to perform well independently, in terms of its compression characteristic, in the 30 – 50 GHz LNA which was fabricated on the same

GaAs wafer – refer to Chapter 5. Despite this 2.1 dB discrepancy, the general consistency between the simulated (Fig. 7-25) and measured characteristics supports the view that the multiplier design is sound and its response is predictable.

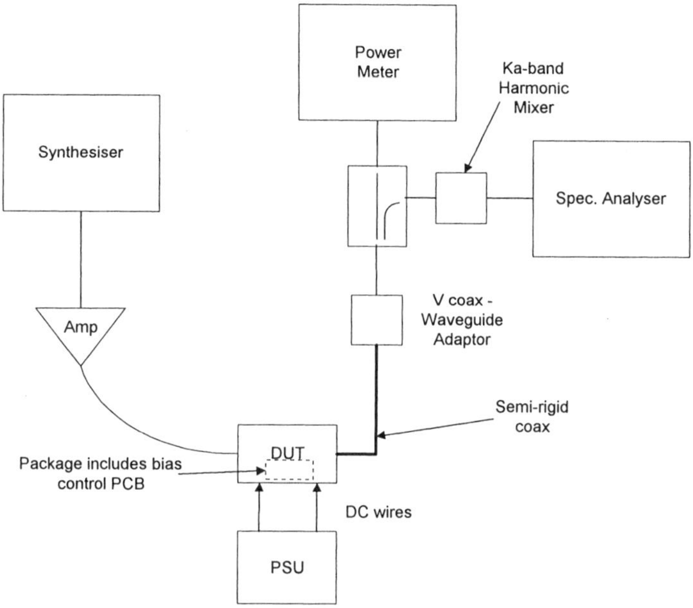

Figure 7-27. Packaged 40 GHz frequency doubler test setup.

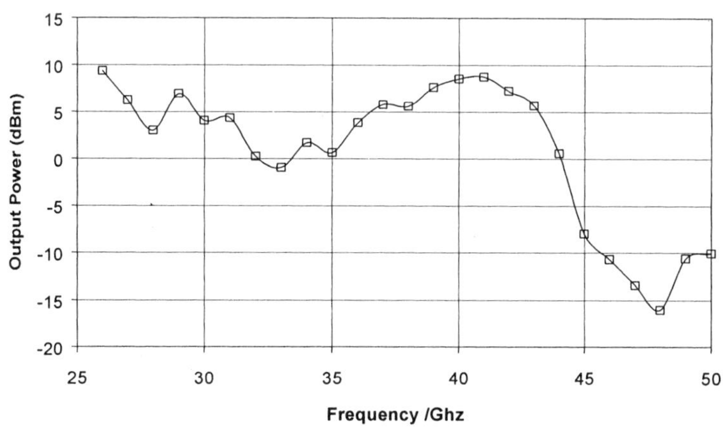

Figure 7-28. Measured output power versus frequency for fabricated doubler for input power of +9 dBm.

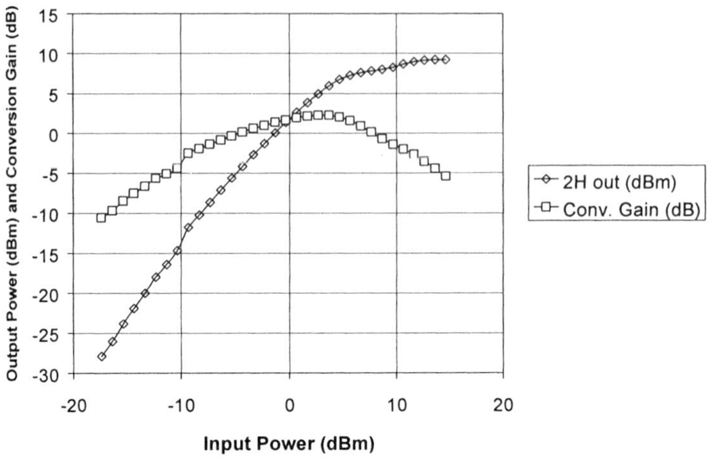

Figure 7-29. Measured second harmonic output level (dBm) and conversion gain (dB) as a function of the incident level (dBm) at 20 GHz.

A contour plot for the measured second harmonic output as a function of drive and bias is presented in Fig. 7-30.

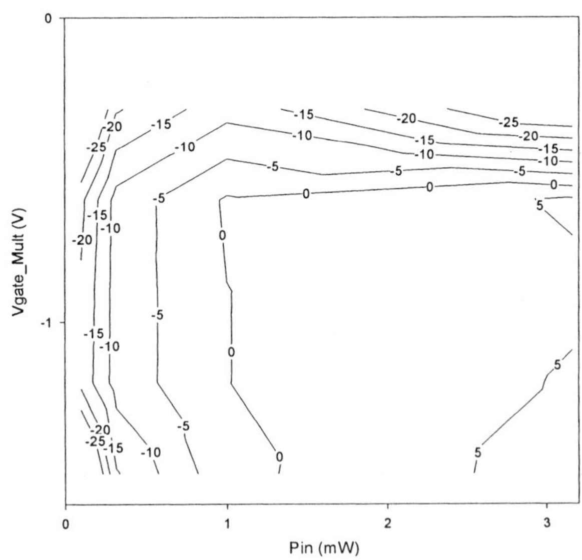

Figure 7-30. Contour plot showing second harmonic output level as a function of multiplier stage gate bias and incident level in mW.

For convenience, the incident power has not been transformed to an equivalent voltage. It is clear that the second harmonic contours are similar in form to those of the generic FET doubler discussed earlier in this chapter – see Fig. 7-6. This comprises additional validation of the theory presented in that section and further validates the design approach adopted for the doubler presented in this section.

3.4 Practical Single-Ended Multiplier Design Conclusions

It is clear that the performance achieved with the 40 GHz doubler is significantly better than in the case of the 56 GHz tripler. There are many factors involved in this performance differential.

1. The higher frequency output requirement from the tripler places additional demands on the high frequency capability of the H40 process. In fact, the output frequency of the tripler is quite close to the f_t of the process.
2. As has been discussed in Section 2 of this chapter, the inherent doubling capability of a FET device is superior to its tripling capacity.
3. The tripler design consisted of a single-stage multiplying device with stubs used for matching and unwanted harmonic rejection. In the case of the doubler, an additional buffer stage of amplification is included. While adding to the circuit complexity, this approach ensures improved power output capability and improved unwanted harmonic filtering by virtue of the frequency response of the amplifying stage.

Given the compact layout of the amplifying stage (see Chapter 5), its use at the output of the multiplier is well justified. It is the authors' strong belief that future mm-wave MMIC single-ended multiplier developments will tend to use such buffer amplifiers for improved performance.

Thus far, the multipliers studied have consisted of a single FET type device. Typically, the device is embedded between matching circuits that also play a filtering role. Such filtering can be limited depending on the desired output harmonic. It has been explained why fundamental filtering is sometimes difficult to achieve, in particular in the case of a frequency tripler. As a consequence, rejection of unwanted harmonics in a single-ended multiplier may not be adequate for an application and other approaches need to be considered. One of these approaches is the combination of two devices in a balanced configuration, where the phasing effects can be exploited to improve the rejection of one or more of the unwanted harmonics. Such a balanced approach can also increase the output power capability of a multiplier and this is of potentially great benefit at mm-wave frequencies where the power output of devices becomes constrained.

4. BALANCED FREQUENCY MULTIPLIER CONSIDERATIONS

4.1 Balanced Frequency Multiplier Introduction

The design of mm-wave MMIC frequency multipliers demands that careful attention be given to unwanted harmonic (especially fundamental) rejection; moreover, at high frequencies, the output power capability of an individual MMIC device can become a constraint. To address these issues, balanced multiplier configurations can be considered[67, 69, 71]. These enhance the power output capability, and at the same time some additional unwanted harmonic rejection may result.

The theory underlying such balanced multiplier configurations is well-understood[66]; the characteristics of the planar hybrids employed in practical MMIC circuits play a key role in their successful implementation. In the following sections, some of the key aspects of balanced multiplier configurations are studied. The section opens with a brief discussion of a generic balanced multiplier configuration, and the defining relations linking the output harmonic characteristics to the hybrid parameters are outlined. Next, the practical implementation of balanced multipliers, particularly in the context of published planar/MMIC designs, is considered. Optimum design strategies for frequency doublers and triplers will be discussed, in particular in terms of the most appropriate hybrids for a given MMIC requirement.

A novel generalised balanced multiplier configuration is then presented. In principle, this approach makes possible a flexible multiplier capable of having its harmonic generation factor configured electronically. The harmonic generation characteristics associated with this multiplier configuration are discussed. This multiplier concept is suited to a wide range of applications, particularly high frequency instrumentation applications where the flexible harmonic number would be of tremendous benefit.

4.2 Balanced Frequency Multiplier Theory

A balanced frequency multiplier generally utilises microwave hybrids in conjunction with a pair of single-ended multiplier circuits. The hybrid-multipliers-hybrid arrangement is configured in such a way that the signals due to each individual multiplier circuit, at the desired output harmonic, *add* in phase at the output port. If the single-ended multipliers are saturated, this approach improves the available output power by up to 3dB. Simultaneously, it is possible that unwanted harmonics may be rejected due to cancellation. In some cases, it is convenient to replace one of the hybrids with a direct connection – see for example Section 4.3.2 of this chapter.

A generalised balanced multiplier schematic is presented in Fig. 7-31. This consists of an input $a°$ hybrid, a pair of identical single-ended multiplier circuits, and an output $b°$ hybrid. In general, the phase factors a and b can be associated with hybrid characteristics alone, but may also include contributions due to unequal path lengths in the multiplier arms between the hybrids and the single-ended circuits. These phase factors can vary significantly with frequency, and let us denote their values at the nth harmonic by a_n and b_n respectively.

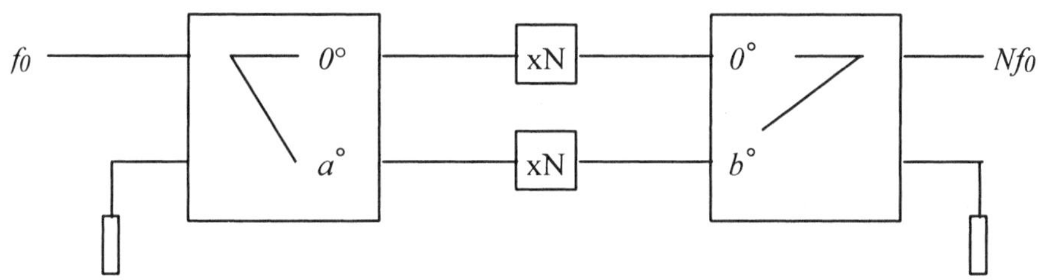

Figure 7-31. Balanced xN frequency multiplier schematic.

In terms of commonly used planar hybrids, the Wilkinson structure[6] has perfect amplitude and phase balance due to symmetry; the Lange coupler[6] can provide an approximately 90° phase difference factor over a broad frequency range (the corresponding amplitude characteristic is not as well balanced); the ratrace (180°)[6] and branchline (90°)[6] structures provide their nominal phase difference factors over relatively narrow bands.

Much has been published on the topic of FET frequency multipliers, including for example works by these authors[72, 75]. Sections 2 and 3 of this chapter have discussed this topic in considerable detail. In simple terms, for a given input drive level, optimum conversion to a given harmonic requires the FET to be biased appropriately. It is usually inevitable that undesirable harmonic outputs will also be generated.

From Fig. 7-31, the following *balanced multiplier design equations* ensue:

$$Na_1° + b_N° = 360° \cdot j$$

$$\Delta\phi_m = ma_1 + b_m$$

$$L_m(dB) = 10\log_{10}\left[\cos^2\left(\frac{\Delta\phi_m}{2}\right)\right]$$

where N is the desired output harmonic number, j is an integer, $\Delta\phi_m$ is the phase difference in the combined outputs at the mth harmonic, and L_m is the rejection provided by the balanced configuration's phase characteristic at the mth harmonic. Contributions to the mth harmonic rejection, due to the hybrid's imperfect amplitude balance characteristic and due to the single-ended circuit rejection properties, are not considered.

4.3 Frequency Doubler Configuration Review

A number of possible planar doubler configurations are outlined and discussed in this section.

4.3.1 $a_1 = 180°/ b_1 = 0°$

With $N = 2$, a solution of the balanced multiplier design equations above is $a_l = 180°$, $b_m = 0°$ ($\forall m$). This configuration results in cancellation of the odd harmonics at the output. The hybrids can be conveniently implemented in MMIC technology as Ratrace and Wilkinson structures respectively. Provided the input hybrid gives the required 180° phase difference at the device inputs, the expected odd harmonic cancellation follows, provided of course that the active devices are matched. The Wilkinson structure frequency response is centred at the second harmonic to maximise the combined output at that frequency. Higher even order harmonics will also add in phase at the output, but the degraded power transmission response of the Wilkinson combiner at these frequencies helps to suppress them.

4.3.2 180°/0° ([67])

Alternatively, the output Wilkinson combiner can be replaced by a direct connection, as reported by Zirath et al[67] for a W-band MMIC doubler. The input 180° phase difference is provided by a Wilkinson splitter with a path length difference of $\lambda/2$ (at the fundamental) in the transmission lines connecting the splitter to the devices in the single-ended multipliers. The resulting MMIC multiplier is very compact. A characteristic of this approach is that the effective load impedance presented to the individual multiplier circuits is 100 Ω, rather than the conventional 50 Ω that would be the case when an output hybrid is used. As a consequence, a single-ended doubler optimised for a 50 Ω load may require some modification to perform optimally in such a direct connection configuration. This implementation's performance is sensitive to the separation between the active elements and the output tapping point. As evidence of the rejection achievable, a fundamental output 26dB below the second harmonic was reported[67], for an input frequency of 44 GHz.

4.3.3 90°/90° ([69])

An interesting doubler configuration is presented by Gilmore[69]. This consists of a pair of Class-C biased FET devices, nominally biased at pinchoff, connected between two Lange coupler hybrids. The analysis presented[69] suggests that a full-wave rectified signal is generated at the combined multiplier output. Such a waveform contains even harmonics only. However, it can be shown that the second harmonic content of this waveform is not optimum - in fact the FET devices would need to be biased beyond pinchoff in order for this condition to be achieved[66, 72]. Furthermore, there is a possible weakness in the analysis presented by Gilmore[69]: when the two half-wave rectified waveforms, which are a quarter of a fundamental cycle out of phase at the outputs of the FETs, are combined via the second Lange coupler, the assumption is made that this coupler shifts one of the waveforms by an additional quarter of a fundamental cycle so that they mesh together forming a full-wave rectified waveform. At a first glance this argument appears valid, but a more detailed study is justified. A Lange coupler usually has a 90° phase characteristic over a broad frequency range. Thus, the harmonic components (at least the low order ones), of one of the half-wave rectified signals will all be shifted by approximately 90° (rather than 180° for the second harmonic, 270° for the third harmonic and so on that would be required to generate the required time delay). This introduces a frequency dispersive effect, which leads to a distorted version of the half-wave signal at the output of the coupler. This then meshes with the other half-wave signal to produce a distorted full-wave rectified waveform. In fact, from the balanced multiplier design equations above, it is clear that the two second harmonic signals being combined at the output of the second Lange coupler are in quadrature, and this introduces an undesirable 3dB loss. A closer study suggests that this configuration is actually suitable for tripler applications – see Section 4.4.1 of this chapter.

4.3.4 90°/90° (Alternative)

An alternative technique for generating an undistorted full-wave rectified output would be to replace the second Lange coupler with a Wilkinson combiner, and to insert a quarter-wavelength (at the fundamental) section of transmission line between one of the FETs and the combiner in order to achieve the necessary 90° fundamental phase shift. This line will correctly introduce the dispersion-free undistorted time delay prior to the meshing in the combiner. It follows that the even harmonics add and the odd harmonics cancel, as in Section 4.3.1 of this chapter. In terms of the phase factors, this approach can be characterised by $a_1 = 90°$, $b_m \approx 90m°$.

4.4 Frequency Tripler Configuration Review

Two possible MMIC tripler configurations are considered in this section.

4.4.1 90°/90°

A solution to the balanced multiplier design equations for $N = 3$ is $a_1 = 90°$, $b_3 = 90°$ (refer to the discussion on[69] in Section 4.3.3 of this chapter). A successful implementation of this approach, using input and output Lange couplers, was reported by Fudem and Niehenke[71] for both Q-band and W-band MMIC triplers. The tripling action[71] is based on over-driven Class-A biased HEMT devices where the second harmonic generated is intrinsically low, and as a result the limited rejection (3dB) due to the balancing is not a major concern. The use of an output Lange coupler, with a broadband phase characteristic of 90° at both the fundamental and third harmonics, facilitates some fundamental rejection. The fundamental rejection that can be achieved in practice may be constrained by the fact that the *amplitude* balance of the Lange structure cannot be maintained over such a broad band. Nevertheless, a fundamental output 40dB below, and a second harmonic output 20dB below the desired third harmonic output have been reported for the Q-band tripler[71].

4.4.2 180°/180°

Another solution of the balanced multiplier design equations for $N = 3$ is given by $a_1 = 180°$, $b_3 = 180°$, which could, for example, be implemented in MMIC technology using two Ratrace structures. In this configuration, if the output hybrid's 180° phase characteristic applies over a broad band, both the fundamental and the third harmonic add in phase at the output, and the second harmonic is rejected. Such a scheme could be useful for a balanced tripler in which the second harmonic cancellation is critical, and in which the fundamental rejection is not a concern, is intrinsic in the single-ended designs, or is achieved by means of appropriate high-pass filtering after the output hybrid. However, these output hybrid harmonic characteristics may be difficult to realise due to the narrow-band nature of the Ratrace structure.

4.5 A Novel Generalised Balanced Frequency Multiplier Approach

In this section, design details on a proposed generic balanced FET multiplier configuration are introduced. The novel design can be configured easily for balancing a pair of single-ended MMIC FET multiplier circuits of any order. This approach makes available the desired 3dB output power improvement and also provides some rejection of all the unwanted harmonics. The generic multiplier schematic is presented in Fig 7-32. The schematic consists of an input Wilkinson hybrid and a pair of FET-based single-ended *xN* multipliers with a direct connection at their output. The path lengths between the input hybrid and the individual multipliers differ by θ_0° at the fundamental centre frequency, f_0, this path length being implemented by an electronic phase shifter. The design equation for optimum conversion to the Nth harmonic is

$$\theta_0 = \frac{360}{N}$$

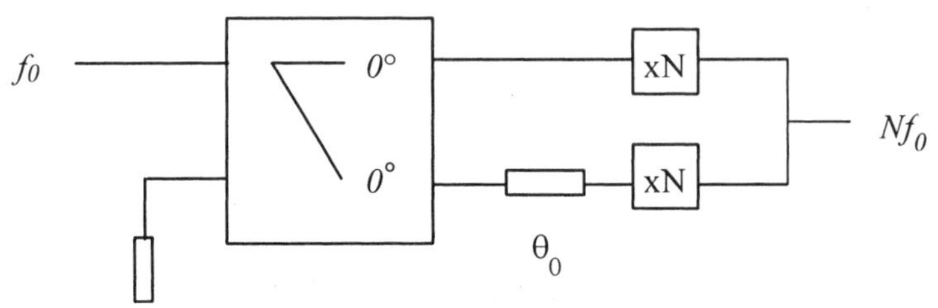

Figure 7-32. Generic balanced multiplier schematic.

and the corresponding rejection of the mth harmonic of f_0, due to the effects of the path length difference alone (neglecting dispersion effects), is given by

$$L_m = 10\log_{10}\left[\cos^2\left(\frac{180m}{N}\right)\right]$$

It can be shown that the 3dB bandwidth of the above configuration, in terms of the path length difference influence on the desired N^{th} harmonic characteristic, is 50%, regardless of the harmonic number N. For $N = 2$, the generic multiplier solution corresponds closely to that in Zirath el al[67] (see Section 4.3.2 of this chapter).

Based on the expression for L_m above, the normalised rejection characteristics for a generic doubler and tripler are presented in Fig. 7-33 and in Fig. 7-34 respectively. Note how the desired (normalised) output harmonic trace is the same in both cases.

Considering FET-based single-ended multipliers, and assuming piece-wise linear FET transconductance characteristics, the analysis in Section 2.3 of this chapter can be extended to derive harmonic generation contour plots for this generic balanced multiplier configuration. No account is taken of any additional filtering which may be built into the single-ended circuits. In accordance with the formulation in Section 2.3 of this chapter, the results are presented as normalised contours of harmonic current in the DC bias (normalised) / fundamental input voltage (normalised) plane.

To facilitate direct comparison with the single-ended contour plots, the voltages applied to the individual FET gates have been reduced by a factor of √2, consistent with the 3dB idealised loss in the input hybrid.

The advantage of this generic balanced multiplier configuration over a conventional approach is that the output harmonic can be controlled electronically (electronic control of the bias point for a given harmonic optimisation and similarly electronic control of the phase shifter), with the increased output power capability (relative to a single-ended implementation) still being available. The conventional approach is, generally speaking, specific for a given desired output harmonic, but of course can be designed to give excellent rejection at some unwanted harmonics.

Another advantage of the generic balanced multiplier is that while it offers increased output power at the desired harmonic (which is variable), at the same time it ensures that the powers at all the unwanted harmonics are not increased to the same extent, leading to an improved harmonic rejection ratio. This follows directly from an examination of the expression for L_m and can be seen graphically in the contour plots in Fig. 7-35 and in Fig. 7-36 for the case of a generic frequency quadrupler ($\theta_0 = 90°$). In Fig. 7-35, contour plots for the 3rd and 4th harmonics for both the generic balanced multiplier and the conventional single-ended multiplier are presented.

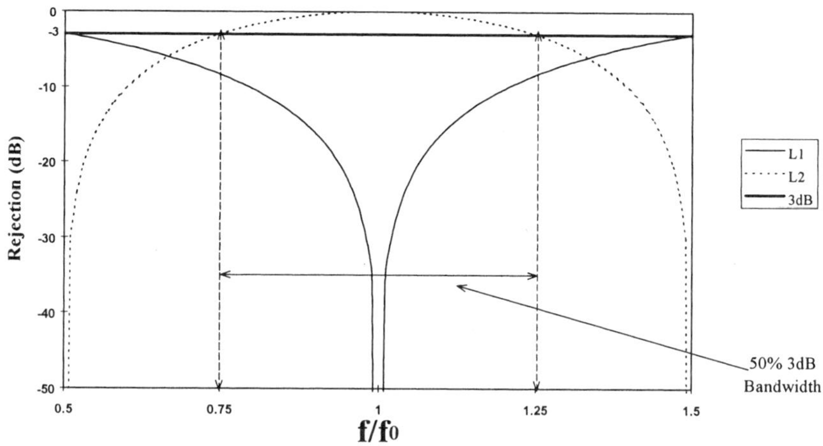

Figure 7-33. Generic balanced doubler rejection characteristics.

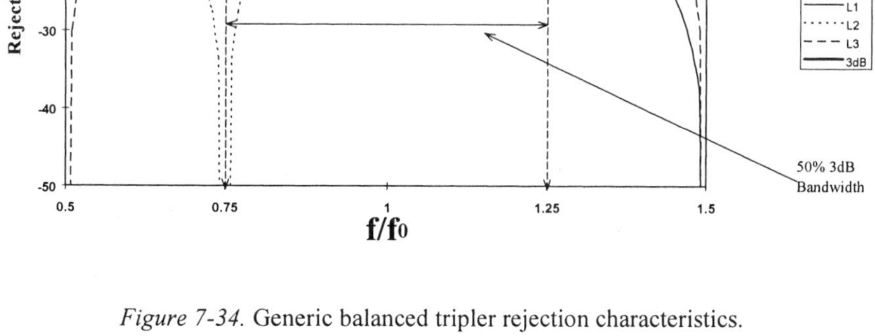

Figure 7-34. Generic balanced tripler rejection characteristics.

When comparing the enhancement of the desired 4th harmonic (when using the balanced rather than the single-ended configuration) and the corresponding enhancement of the 3rd harmonic, one is precluded from simply dividing the corresponding contour levels due to divide by zero difficulties. Instead, a cross multiplication technique has been developed by the authors, yielding a useful figure of merit for the balanced multiplier, as outlined in Fig. 7-35. This technique can be understood from the following:

$$4H\ Enhancement : E_{4H} = \frac{a_{4Balanced}}{a_{4SingleEnded}}$$

$$3H\ Enhancement : E_{3H} = \frac{a_{3Balanced}}{a_{3SingleEnded}}$$

$$Relative\ Enhancement : E_{rel} = E_{4H} - E_{3H}$$

In order to eliminate divide by zero issues, multiply both sides by $a_{4SingleEnded} \cdot a_{3SingleEnded}$ to yield the Generic Multiplier Figure of Merit (GMFM):

$$GMFM = a_{4Balanced} \cdot a_{3SingleEnded} - a_{3Balanced} \cdot a_{4SingleEnded}$$

The GMFM represents, qualitatively, the difference between the relative increases (over the single-ended case) in the fourth harmonic and third harmonic. The GMFM contours, derived from the individual plots in Fig. 7-35 are shown in Fig. 7-36. Positive GMFM values represent regions in the bias/drive plane where the fourth harmonic is enhanced more than the third with respect to the single-ended circuit. It is easily seen by inspection in Fig. 7-35 and in Fig. 7-36 that the region in Fig. 7-36 where the GMFM is greatest is also the region where optimum fourth harmonic output is expected. Similar GMFM contours can be derived for all the other harmonics, as appropriate.

4.6 Balanced Frequency Multiplier Recommendations

For requirements where the unwanted harmonic rejection properties of a given configuration are of concern, it is advisable to avoid the use of output hybrids whose phase characteristics are highly sensitive to frequency. This is due to the fact that the phase characteristics at the harmonic frequencies could be difficult to predict accurately and, as a result, the phase cancellation effects would be difficult to realise, particularly in volume, and would probably be narrow-band. As a consequence, the broadband Wilkinson and Lange couplers are the preferred output hybrids for MMIC frequency multipliers. Furthermore, in order that excellent unwanted harmonic cancellation is achieved, the output hybrid is also required to have a good amplitude balance at the various harmonics; the symmetric Wilkinson structure meets these twin criteria.

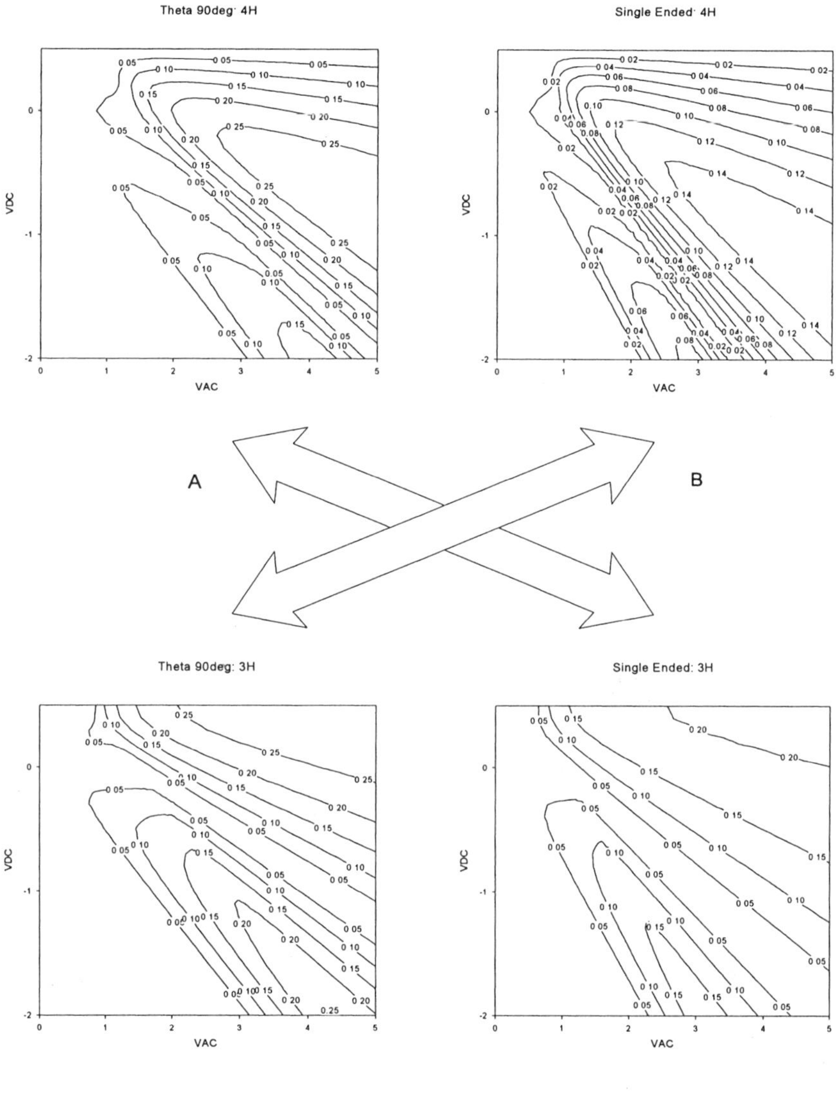

Figure 7-35. Development of the Generic Multiplier Figure of Merit (GMFM) for $\theta_0 = 90°$ Balanced Generic Quadrupler (4th harmonic relative to 3rd harmonic output enhancement) . GMFM is defined as A – B, where A is the product of the contour plots for the Balanced Generic Multiplier's 4th harmonic and the conventional Single-ended Multiplier's 3rd harmonic, and B is the product of the contour plots for the Balanced Multiplier's 3rd harmonic and the conventional Single-ended Multiplier's 4th harmonic.

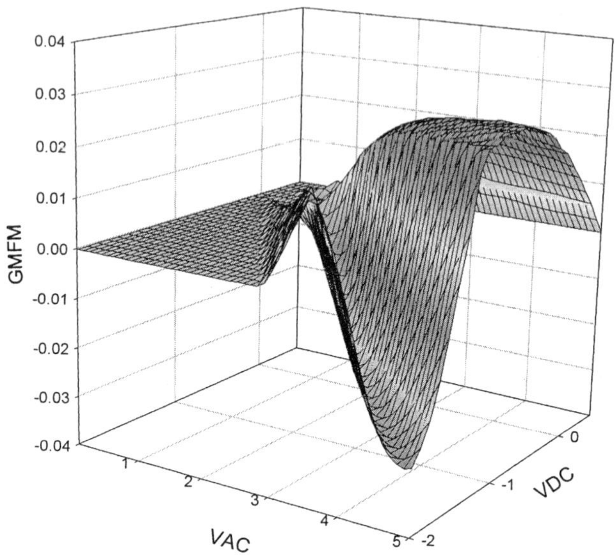

Figure 7-36. Generic Multiplier Figure of Merit 3D contour plot for the case of the 4th and 3rd harmonic outputs from a $\theta_0 = 90°$ generic balanced multiplier.

The frequency sensitivity of the input hybrid does not play a significant role in the performance of narrow-band balanced multipliers, provided of course that it introduces the required phase difference at the active element inputs. However, the response of this hybrid does impact directly on the potential bandwidth of the multiplier. As a consequence, broadband MMIC multipliers generally avoid the use of relatively narrow-band input hybrids, such as Ratrace and Branchline structures. In most situations, the harmonic selectivity properties of a balanced multiplier require that the two single-ended circuits are driven by out of phase signals. This generally precludes the use of the Wilkinson splitter as the input element, unless unequal lengths of transmission line to the active elements are used to introduce the phase difference as discussed in Section 4.3.4 of this chapter.

In terms of multipliers with a fixed harmonic output, as discussed above, the recommended broadband tripler approach would be that proposed in[71], as outlined in Section 4.4.1 of this chapter. This is the preferred MMIC configuration due to the use of the broadband Lange couplers.

Recommending a broadband MMIC doubler approach is not as straightforward because the various options discussed include frequency sensitive components (either in the form of narrow-band hybrids or lengths of transmission line). A good approach to adopt is that suggested as an improvement on[69] in Section 4.3.3 of this chapter; the frequency sensitive component there is only 90° in length at the fundamental, whereas the frequency sensitive components associated with the other options are 180° in length. This shorter length reduces the overall frequency sensitivity of the design.

The generic configuration outlined in Section 4.5 of this chapter would be attractive for applications, such as instrumentation, that require varying harmonic generation factors. While this approach does not necessarily offer the same specific unwanted harmonic rejection characteristics as, for example, a dedicated doubler or tripler implementation, its non-specific harmonic generation factor certainly makes it an interesting topic for further study.

4.7 Balanced Frequency Multiplier Conclusions

A number of dedicated planar frequency doubler and tripler configurations have been assessed and relative tradeoffs in terms of unwanted harmonic rejection have been analysed. Conclusions have been drawn concerning the best approach to adopt for a given requirement, in particular for broadband applications.

A novel generic balanced MMIC multiplier configuration has been introduced. The bandwidth characteristic due to its intrinsic path length difference is 50%. This generic design may not provide as much unwanted harmonic rejection as some of the other design configurations discussed. However, its simple structure makes it attractive for applications where such harmonic rejection is not an issue, or where the harmonic rejection can be provided by other means. Moreover, by employing an electronic phase shifter as the θ_0° transmission line section, the multiplication factor can be electronically controlled, resulting in a circuit of potential interest in a wide range of applications.

5. HIGH POWER GENERATION AT MM-WAVE FREQUENCIES USING FET MULTIPLIERS

5.1 High Frequency High Power Generation Introduction

Thus far, the theoretical and practical aspects underlying the operation of a FET frequency multiplier have been considered in detail, and the extension of this to a balanced configuration has also been discussed. This section is intended as a tutorial covering the key issues to be considered in the actual development of a high power high frequency source based on a MMIC frequency multiplier chain.

The key objective of this study was to assess the suitability of GaAs MMIC technology for the generation of a significant level of mm-wave power. With the rapid development of GaAs processes in recent years, significant progress has been reported in terms of monolithic mm-wave circuits. The f_t of a typical 0.25 μm GaAs pHEMT process is now well in excess of 60 GHz, with state of the art processes having f_t values above 100 GHz. As a consequence, monolithic mm-wave frequency multipliers are now feasible – for example, W and Q band multiplier designs are described by Fudem and Niehenke[71].

As a nominal goal for this section's study, an output power of 100mW at 100 GHz was targeted with an input of 100mW at a relatively low frequency (25 GHz or 16.667 GHz) being available. Such a goal is highly aggressive in light of the current state of the art, and also represents a reasonable target to make such a solution of potential interest for emerging high frequency applications.

The study was simulation based, and in common with all the other simulations described throughout this work, the high frequency harmonic balance CAE tool, HP-Eesof's Libra[27], was used.

The primary output of this section is an assessment of the art of the possible[76]. This analysis has stimulated further effort in this area, with a full-scale wafer design, fabrication and evaluation now underway[77].

5.2 Current State of the Art Review

A literature review was carried out and a number of relevant papers were identified. In Huang et al[78], details are presented of a state of the art 94 GHz power amplifier. Both Zirath et al[67] and Fudem and Niehenke[71] provide interesting information on W-band frequency multipliers. In[67], details are presented of a balanced doubler to 94 GHz with a conversion loss of 15dB. In[71], brief details are outlined for a frequency tripler to W-band which generates +5.5 dBm at 99.5 GHz (the input power is not specified). Other relevant papers involved a varactor based MMIC W-band frequency tripler[79] and a 1W W-band transmitter[80]. Both of these designs are capable of delivering high power levels at W-band.

5.3 Potential MMIC Multiplier Schemes

In order to generate power at 100 GHz, a number of options exist in terms of the multiplier configuration. MMIC multipliers with specific varactor devices integrated have been reported[79]. However, standard MMIC processes do not typically support efficient varactor devices. For the development of a MMIC multiplier chain, it would certainly be prudent to make use of standard MMIC process technologies in which the conventional non-linear element is a MESFET or one of its HEMT variants. Such an approach would result in a design that is not overly constrained in terms of being inextricably dependent on a unique fabrication process.

As outlined in Section 2 of this chapter, unlike varactor based frequency multipliers, FET type multipliers are generally restricted to low order multiplication factors. FET doublers and triplers are commonly employed, as higher order multipliers are less prevalent due to the degraded conversion efficiencies that are achievable[66, 72]. Stenger et al[80] reported on an interesting times-six multiplier chain, delivering an output power of 1W in W-band. This design makes extensive use of MMIC technology, but a number of distinct chips are involved, thereby leading to a very complex assembly operation. The package machining is also very difficult. Moreover, it appears from[80] that MMICs from different foundry processes are involved. The focus of the work described in this study was to put as much as possible, if not all, of the chain design on a single GaAs chip, thereby resulting in a simpler assembly process. Of course, yield issues for very large circuit areas would also demand consideration.

On this basis, a number of potential multiplier chain configurations were identified at the outset. The aim was to study these schemes in some detail to assess their suitability for the intended application. These chains are summarised in Table 7-3.

Table 7-3. Multiplier chain configurations studied

Input Frequency (GHz)	Chain Detail	Output Frequency (GHz)
25	Doubler/Doubler (X2X2)	100
16.667	Tripler/Doubler (X3X2)	100
16.667	Doubler/Tripler (X2X3)	100

In order to maintain appropriate power levels through the chains, it was envisaged that amplifiers would also be required as appropriate at the multiplier outputs.

5.4 Non-Linear mm-wave Device Models

To make the simulation study meaningful, a valid non-linear device model was required. A paper containing extensive detail on a state of the art high frequency device was reviewed[78]. This paper describes a high power W-band amplifier developed using TRW's 0.1um GaAs pHEMT process. The process f_t is 130 GHz, and f_{max} is 200 GHz. Non-linear model data, suitable for use with the Curtice-Ettenberg equivalent circuit model, are presented in this paper for an 8x20 μm device. This is the same device that TRW used in the 94 GHz amplifier development. In order to achieve the 350mW output power level, a number of 8x20 μm device based MMICs were combined in a module. The individual two-stage MMICs delivered a maximum 300mW of power with a linear gain of 8 dB; in fact the output amplifier stage essentially consisted of 16 of these 8x20 μm devices in parallel. For the purposes of this study, the significant aspects of the TRW paper are twofold:

1. High power capability near 100 GHz has been demonstrated
2. A valid non-linear (large-signal) model is detailed

Using the information provided in[78], a model of the TRW 8x20um device was developed within the framework of the Libra simulation tool. The model validity was verified by means of comparisons between simulation responses and the corresponding measured responses published in[78]. This model is considered by the authors a reasonable representation of a state of the art high frequency device with good power output capability. The non-linear model as presented in[78] is shown in Fig. 7-37.

5.5 Non-Linear Building Block Circuit Designs

Some building block frequency multiplier and amplifier circuit schematic designs were developed using the model in Fig. 7-37. The designs consist of the non-linear devices, with associated input and output matching network transmission lines being based on the microstrip library models for the GMMT H40 foundry process. This results in simulations that are as realistic as possible. No attempt was made to lay out the circuits as this was outside the scope of this study, and would require a complete set of models from a specific foundry.

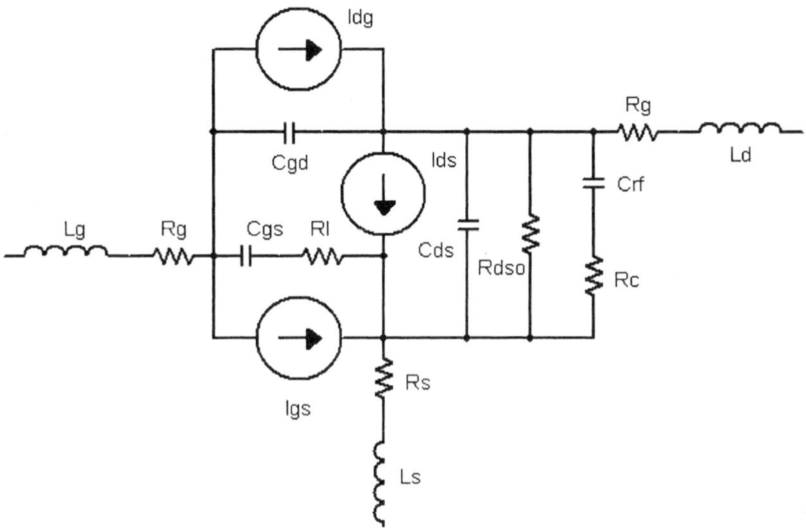

a_0 = 0.4309E-01 A, a_1 = 0.915E-1 A/V, a_2 = 0.2149E-1 A/V^2,
a_3 = -0.2996E-1 A/V^3, β = 0.3684E-1 V^{-1}, γ = 2.1081 V^{-1}, Rdso = 296 Ω

Figure 7-37. Curtice-Ettenberg equivalent circuit model for TRW 8x20 um device (source Huang et al[78], © 2004 IEEE).

All the building blocks are single-ended single-stage designs, consisting of a 8x20 um device embedded in appropriate matching networks. The circuits were designed using standard techniques, and the CAE tool optimiser was applied to refine the first-order designs for optimum performance in the context of output power at the desired frequency. The amplifier designs included a single stub at both the input and output for matching. In the case of the frequency doublers, a pair of stubs was employed at both the input and the output. These stubs are used for both matching and for termination of unwanted harmonics. Similarly, the frequency triplers utilised three stubs at both input and output of the active device. The building blocks developed consist of the following:

- Amplifiers : 33 GHz, 50 GHz, 100 GHz
- Doublers: Input frequencies 16.667 GHz, 25 GHz, 50 GHz
- Triplers: Input frequencies 16.667 GHz, 33.333 GHz

The optimised schematics are presented in Figs. 7-38 through 7-45.

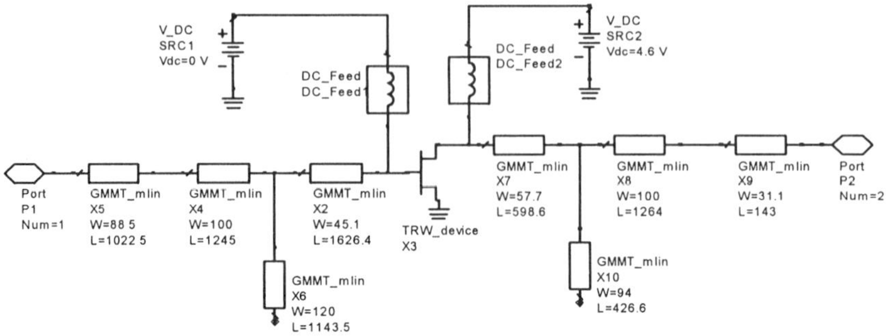

Figure 7-38. Schematic of 33 GHz amplifier.

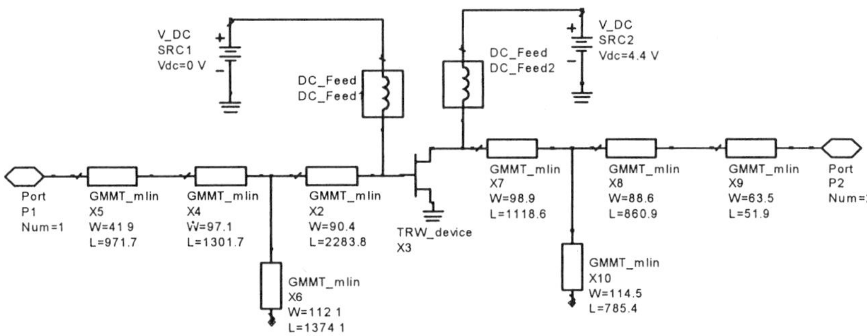

Figure 7-39. Schematic of 50 GHz amplifier.

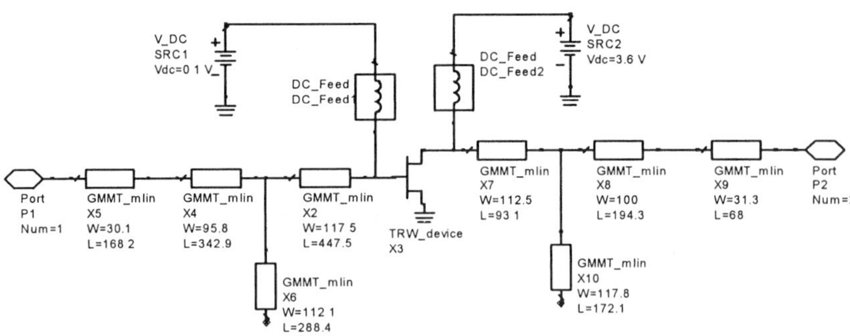

Figure 7-40. Schematic of 100 GHz amplifier.

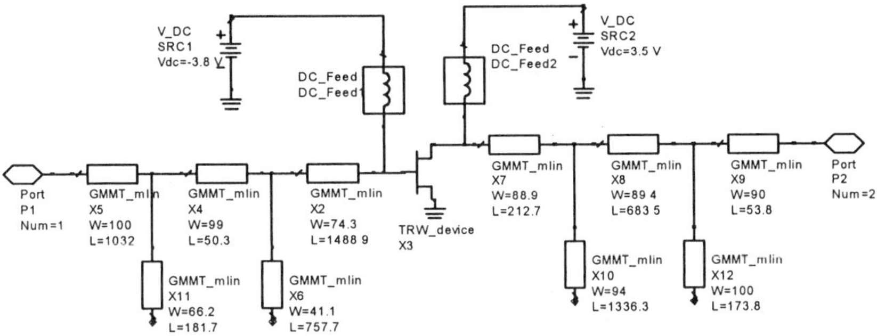

Figure 7-41. Schematic of 16.667 GHz doubler.

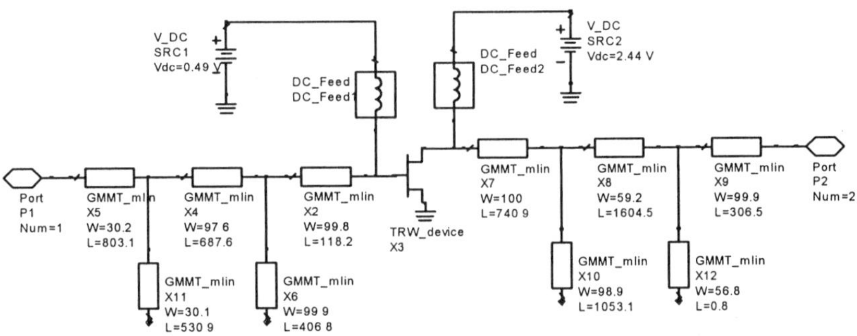

Figure 7-42. Schematic of 25 GHz doubler.

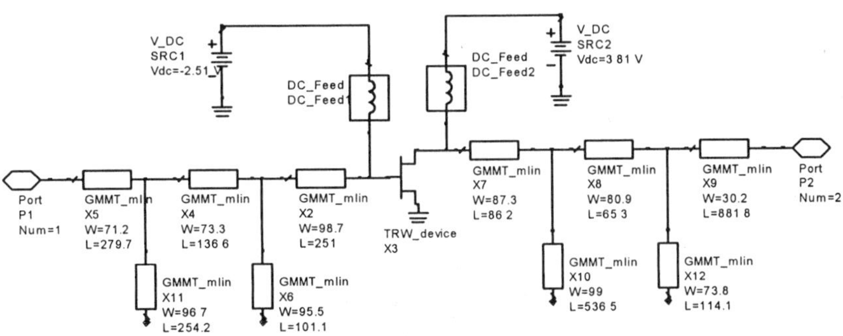

Figure 7-43. Schematic of 50GHz doubler.

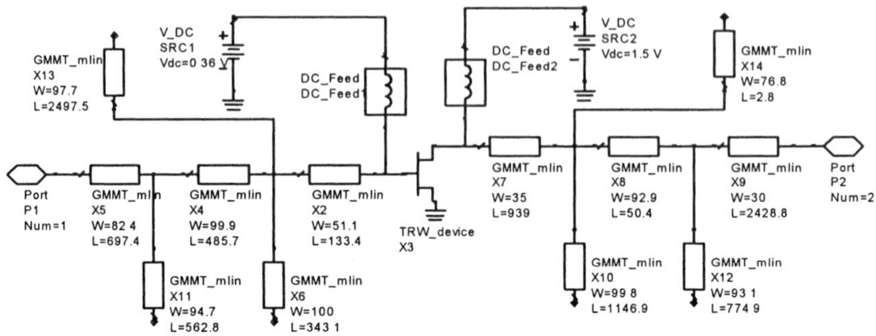

Figure 7-44. Schematic of 16.667 GHz tripler.

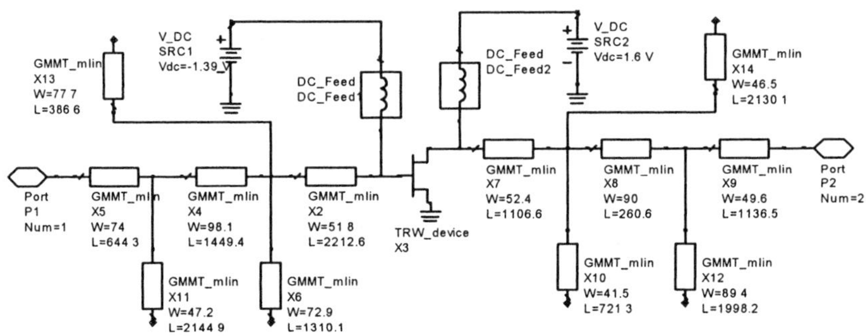

Figure 7-45. Schematic of 33.333 GHz tripler.

The simulated response of our 33 GHz amplifier is presented in Fig. 7-46. The corresponding responses for the 50 GHz and 100 GHz amplifiers are shown in Fig. 7-47 and in Fig. 7-48. Also included in Fig. 7-47 and in Fig. 7-48 are the second and third harmonic outputs, showing the trends in these harmonic levels as a function of input drive. Note that at +20dBm input power, all three amplifiers are saturated.

The predicted responses for 16.6 – 33.3 GHz, 25 – 50 GHz, and 50 – 100 GHz doublers are presented in Fig. 7-49 through Fig. 7-51. The indicated conversion loss of 5 dB for an applied +20dBm at 25 GHz (for the case of the 25 – 50 GHz doubler in Fig. 7-50) is promising and compares well with typical efficiencies of varactor based approaches[66]. Note also that the 16.6 – 33.3 GHz doubler has a poorer conversion response than that indicated by the 25 – 50 GHz simulation – it is believed that this is primarily due to the optimum transmission line lengths for the former exceeding the constraints imposed in the optimisation. It is reasonable that such constraints be imposed, as excessive line lengths (consistent with lower frequency operation) are not viable for MMIC implementation. Not surprisingly, the harmonic performance of the doubler, shown in Fig. 7-51, is poorer than that of the lower frequency designs.

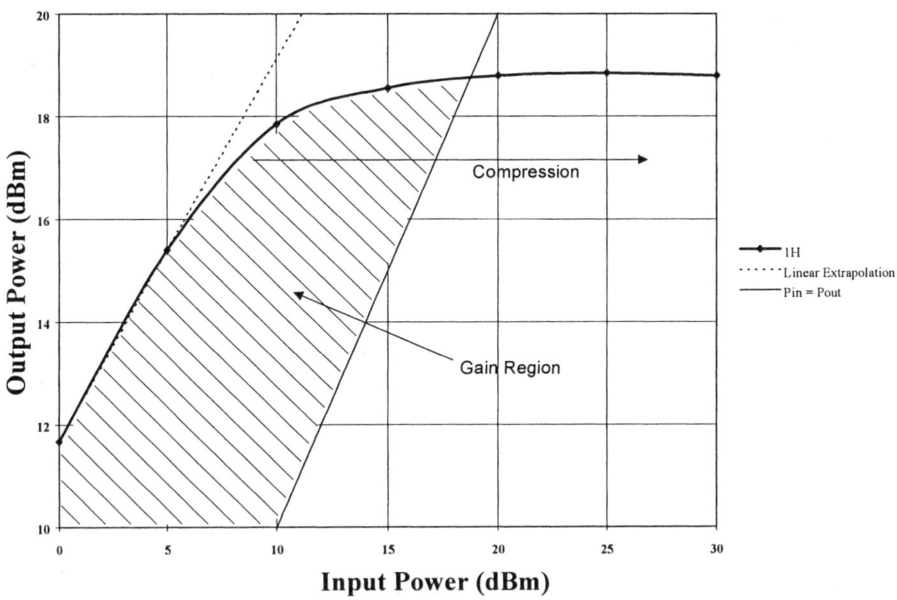

Figure 7-46. 33.3 GHz amplifier response.

The simulated responses of the frequency triplers are shown in Fig. 7-52 and in Fig. 7-53. As expected, the conversion losses of triplers are significantly larger than for similar doublers; moreover, tripler performance degrades more rapidly with increasing frequency. In the case of the 33 – 100 GHz simulation, an input signal of +20 dBm at 33 GHz is converted to +5 dBm at 100 GHz. The harmonic characteristics of the low frequency tripler are also included in Fig. 7-52.

5.6 Chain Responses

The building blocks required for the assessment of the chain configurations defined earlier having been developed, the simulations of the actual chains were next carried out.

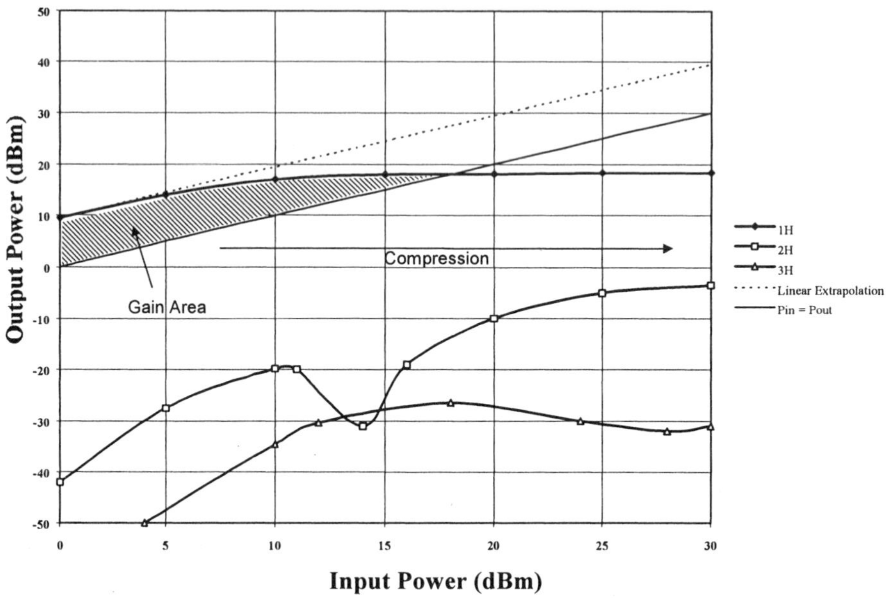

Figure 7-47. 50 GHz amplifier response.

5.6.1 X2X2 Chain

This chain consists of a pair of doublers and a pair of amplifiers, as shown in Fig. 7-54. The output of this circuit when the input power level is swept is presented in Fig. 7-55. The simulated output level at 100 GHz is +16.5dBm for an applied +20dBm. This would be a very useful result if it could be realised. An interesting feature of the response is that the output level is predicted to be +15dBm for an applied level of +10dBm. In other words, the input level can be backed off significantly without degrading the output level excessively.

5.6.2 Other Chains and Discussion

In a similar fashion, the other chain configurations (tripler- doubler and doubler-tripler) were studied. The predicted performances achievable with all three chains are compared in Table 7-4. Based on these results, it appears that the X2X2 chain is the optimum configuration from an output power perspective at 100 GHz. Moreover, the design of a frequency doubler is less complex than a tripler due to less sensitivity to the termination of unwanted harmonics.

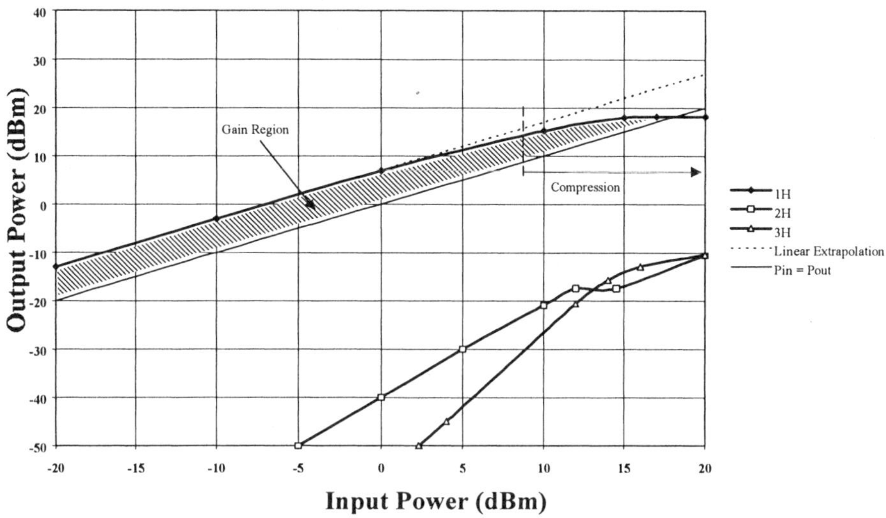

Figure 7-48. 100 GHz amplifier response.

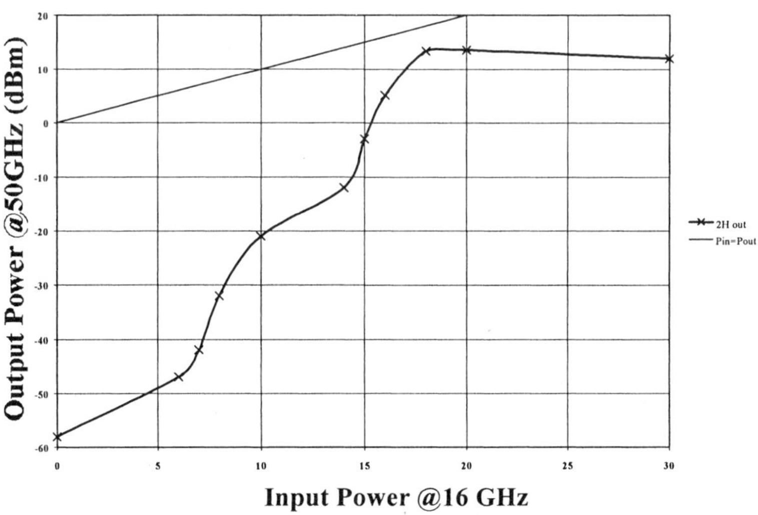

Figure 7-49. 16.6 – 33.3 GHz doubler response.

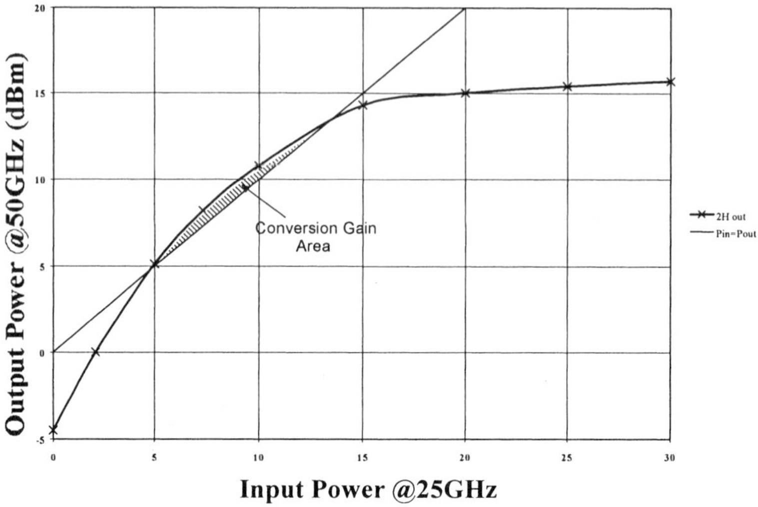

Figure 7-50. 25 – 50 GHz doubler response.

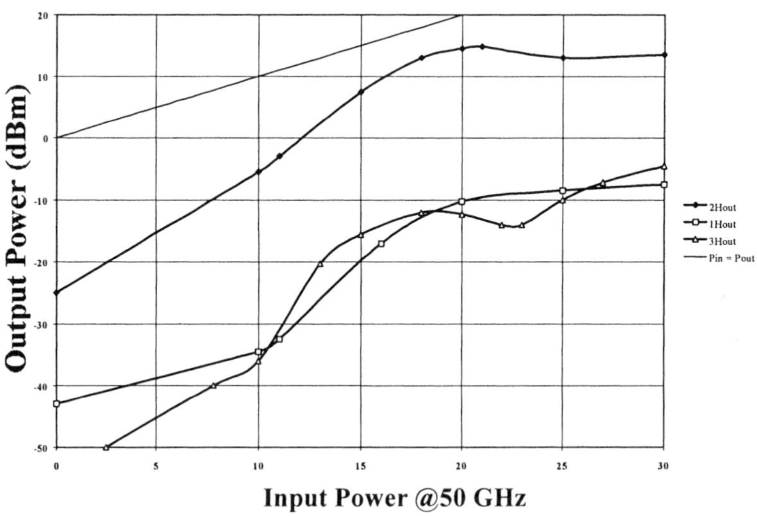

Figure 7-51. 50 – 100 GHz doubler response.

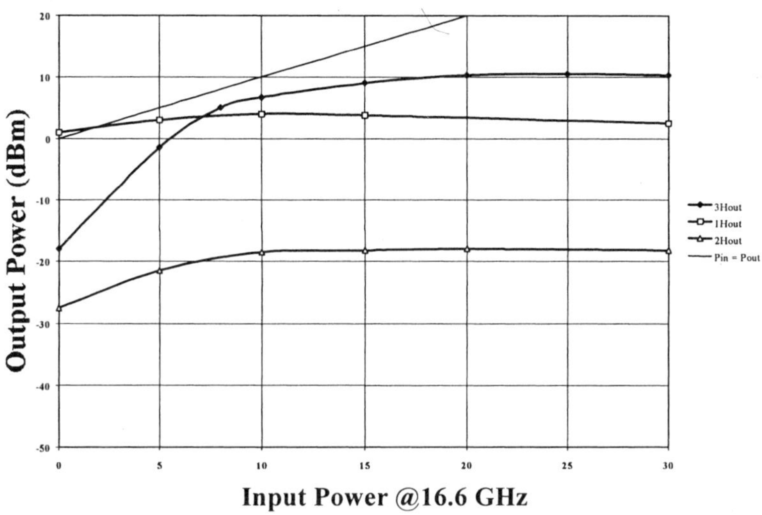

Figure 7-52. 16.6 – 50 GHz tripler response.

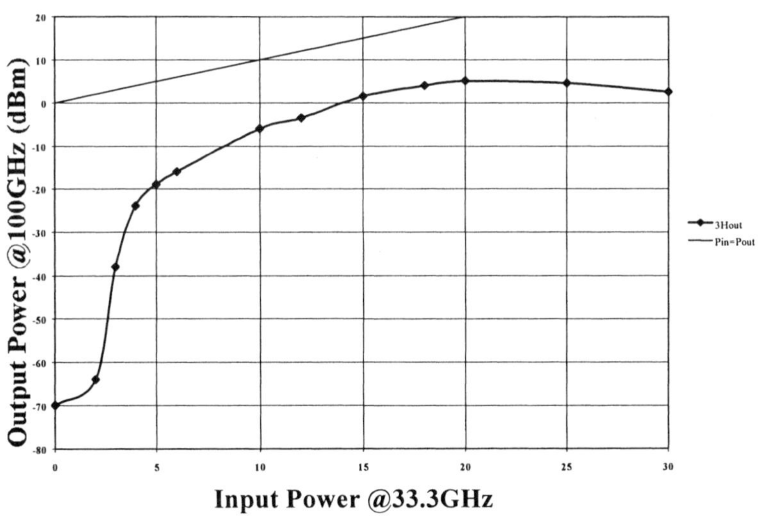

Figure 7-53. 33.3 – 100 GHz tripler response.

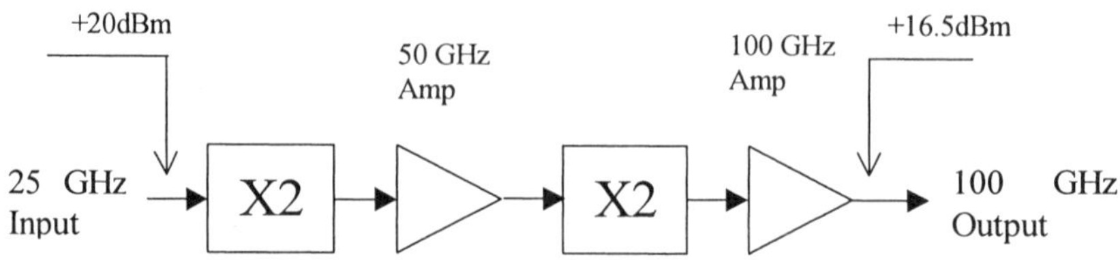

Figure 7-54. X2X2 Chain configuration.

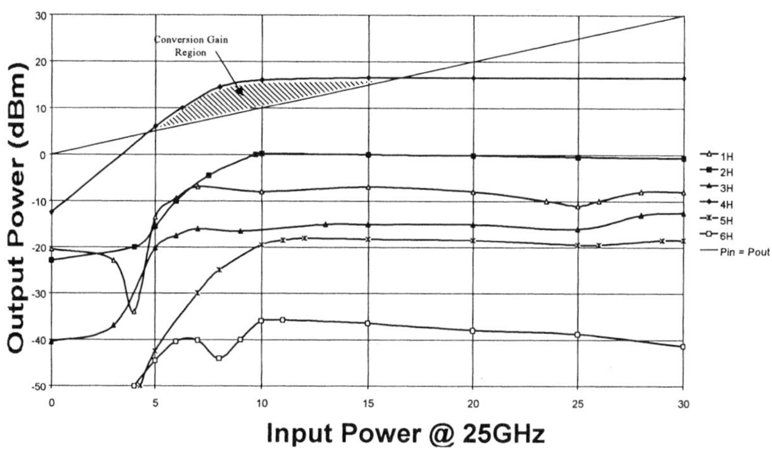

Figure 7-55. X2X2 Single-ended chain response.

Table 7-4. Chain summary

Chain	Input Freq. (GHz)	Input Power (dBm)	Output Power (dBm)	Input Power (dBm)	Output Power (dBm)
X2X2	25	+20	+16.5	+10	+15
X3X2	16.6	+20	+14.5	+10	+11.5
X2X3	16.6	+20	+9.1	+10	<-70

A frequency tripler also has the potential disadvantage that rejection of the fundamental may not be trivial (the conventional quarter wavelength open circuit stub rejection technique is inadvisable as that would also reject the third harmonic!). It has also been found that a tripler may be more prone to severe threshold effects (note the poor performance indicated in Table 7-4 for the X2X3 chain for an input level of +10dBm). On this basis, it was concluded that the X2X2 chain was the configuration most suitable for more detailed examination.

Given that the nominal target was +20dBm at 100GHz, it is clear that the performance of the X2X2 configuration in Fig. 7-55 is still inadequate. An additional power gain of at least 3.5dB is necessary. In order to achieve this, more complex circuit arrangements were considered. As discussed in detail in Section 4 of this chapter, a common technique applied to increase the power output capability of non-linear RF circuits is the application of some form of balancing (paralleling) arrangement using hybrid circuits. Balancing options, in the context of the 100 GHz requirement, are now studied.

5.7 Optimum X2X2 Chain Development

In attempting to develop an effective X2X2 chain, a number of avenues were initially considered. A flowchart detailing the key steps carried out in this phase of the simulation study is presented in Fig. 7-56. Some of the approaches considered and investigated were found not to be appropriate for the ultimate requirement, and were discontinued. In Fig. 7-56, the steps pertinent to the final optimum solution are shown in shadowed boxes.

An interesting method of determining a suitable circuit implementation for the X2X2 chain, that is capable of delivering at least 100mW of power at 100 GHz for an incident level of 100mW at 25 GHz, consists of starting at the output and working back through the chain, making use of the information already compiled for the individual single-ended circuits. This systematic approach is shown to be highly effective in identifying the optimum chain configuration.

Considering the 100 GHz amplifier characteristic, it appears that the most convenient way of delivering at least +20dBm at the output is to have a four-way balanced amplifier. The characteristic in Fig. 7-48 suggests that +16dBm is delivered for an input of +10dBm. By using four such circuits in parallel, and ignoring losses, an output of +22dBm would be expected for a total input +16dBm.

In order to deliver +16dBm from the 50 $\rightarrow$ 100 GHz doubler, a level of +23dBm input into a two-way balanced network would be required; this would result in about +20dBm into each of the single-ended circuits, with a resulting output of +14dBm from each of these circuits, or +17dBm from the balanced network.

To deliver +23dBm from the 50 GHz amplifier, a four-way balanced arrangement is required. By driving the single-ended 50 GHz amplifier with +10 dBm, an output of +17dBm can be generated. A combination of four of these circuits in parallel requires an input of about +16dBm to deliver about +23dBm at the output.

In order to provide +16dBm at the output of the 25 → 50 GHz doubler, a two-way balanced approach is appropriate. By driving the single-ended 25 – 50 GHz doubler with +15dBm, an output of +14dBm is generated. Connecting two of these in a parallel arrangement results in an output of +17dBm for an input of +18dBm. Thus, the nominal +20dBm input ought to be more than adequate to deliver the necessary +16dBm.

For the purposes of this simulation study, these balanced networks have been implemented using Wilkinson structures throughout, centred appropriately at the various frequencies through the chain. If unwanted harmonic rejection was a concern, the doublers could be based on other hybrid approaches (particularly Lange Couplers due to their broadband characteristic – see Section 4.6 of this chapter). However, in this case, the primary concern is the output power at 100 GHz, and the Wilkinson structures are perfectly adequate to assess the potential in that regard. The resulting block schematic is summarised in Fig. 7-57. The large rectangular boxes represent the Wilkinson hybrids. The smaller boxes represent the single-ended doublers, and the small triangles represent the single-ended amplifiers.

Note that three pairs of redundant Wilkinson hybrids have been identified. As well as saving a significant amount of circuit area if the circuit were to be laid out, the elimination of these Wilkinson structures also eliminates the non-ideal losses associated with them. This chain configuration, with the redundant hybrids removed (see Fig. 7-58), has been simulated, and the response is presented in Fig. 7-59. The response is very appealing. Not alone is the output power predicted to be +22dBm at 100 GHz for an input of +20dBm at 25 GHz, but the plot also suggests that there is significant margin in the circuit response, with an output of +20dBm being predicted for an input as low as +10dBm.

Two variations of the above circuit, where the schematic was made more symmetric by making the doubler stages 4-way rather than 2-way balanced, were considered. The block schematics are presented in Figs. 7-60 and 7-61, and the corresponding simulated 100 GHz output power responses are compared with the response of the optimum configuration in Fig. 7-62. It can be seen that these variations degraded the chain response, not alone in terms of the output power which could be achieved for an input of +20dBm, but particularly in terms of inferior performance when the input power was backed off to say +10dBm. This strongly indicates that the approach outlined above is the optimum one for this application. This degradation is consistent with a fundamental understanding of multiplier circuits suggesting that increasing the level of balancing in a multiplier circuit offers superior performance only for relatively high input powers where compression occurs. For lower drive levels, poorer performance is to be expected from multipliers with additional balancing as the input power is reduced.

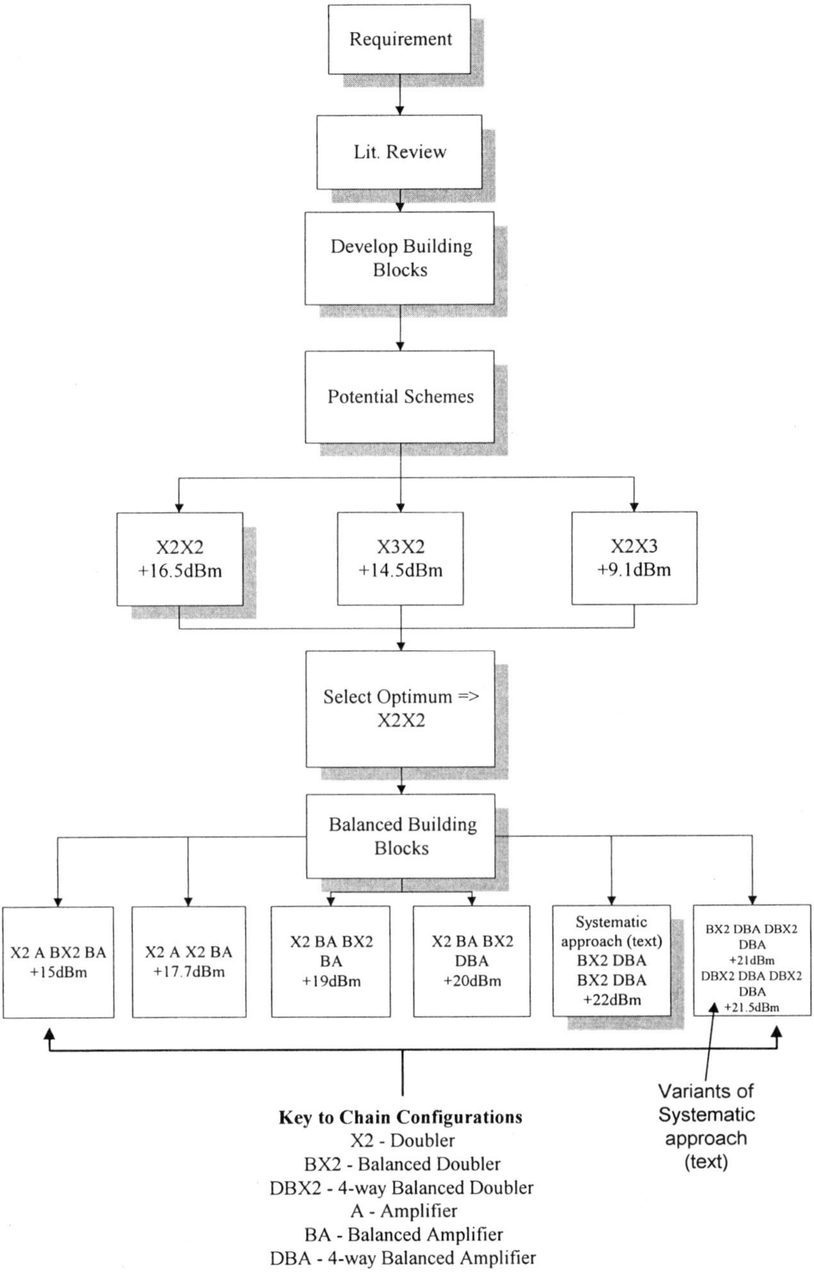

Figure 7-56. Development study flowchart.

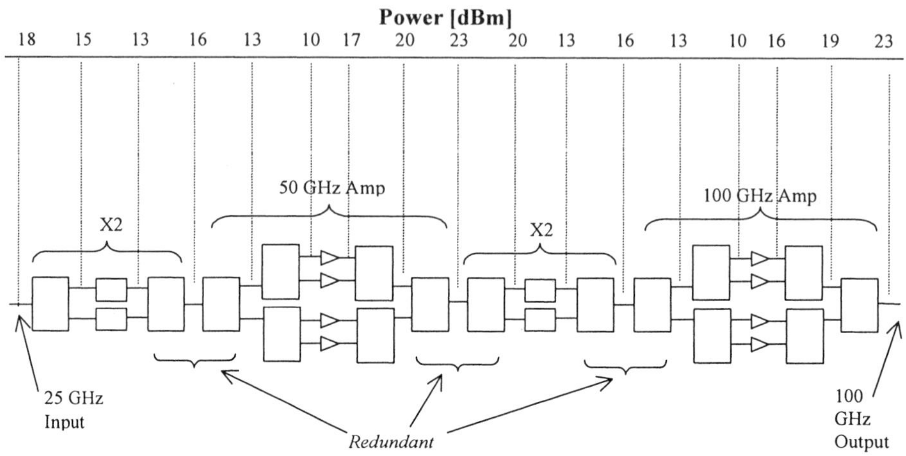

Figure 7-57. Optimum X2X2 chain configuration.

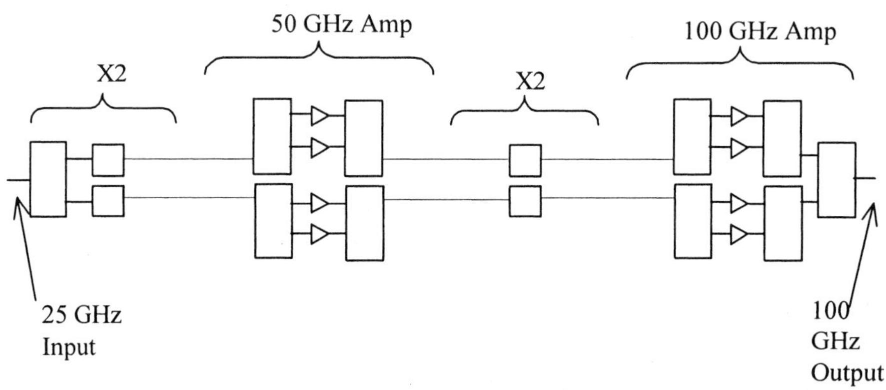

Figure 7-58. Reduced optimum X2X2 chain configuration.

5.8 High Power mm-wave Generation Analysis Conclusions

An investigation into the potential for MMIC based high frequency mm-wave multiplier chains has been carried out. In the course of this work, a systematic approach to the development of an optimum strategy for the cascading of amplifier and multiplier building blocks has been presented.

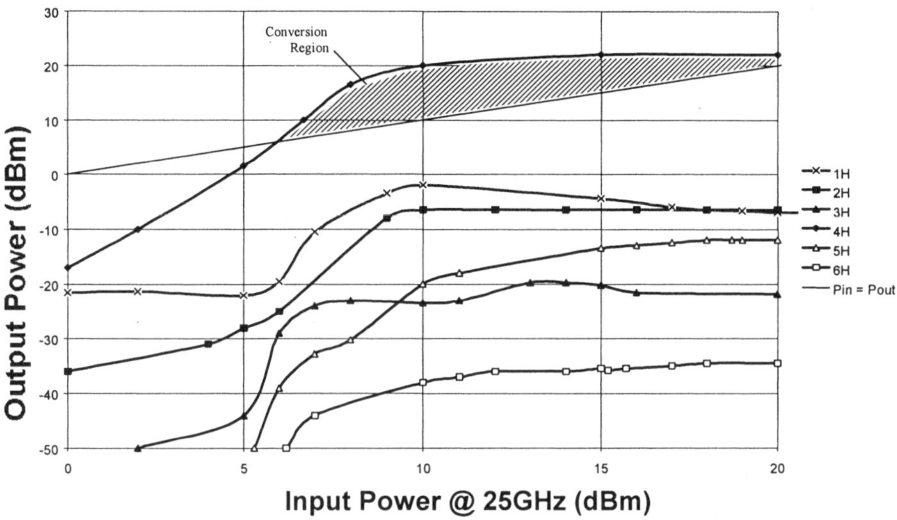

Figure 7-59. Optimum balanced X2X2 chain response.

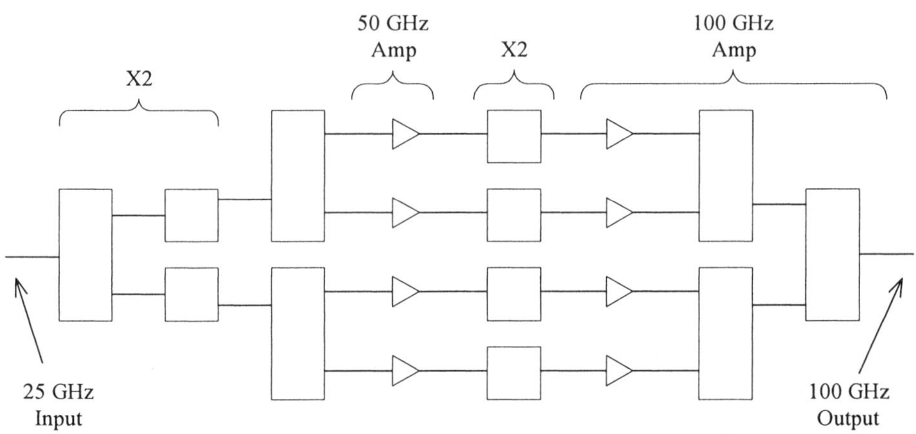

Figure 7-60. First alternative X2X2 chain configuration (Alt#1).

The response in Fig. 7-59 indicates that the 100 mW at 100 GHz goal ought to be achievable using an existing GaAs MMIC process. A conclusive decision concerning whether or not such a design should be fabricated as a single die, or as a number of separate die with more complex packaging requirements, would require a more detailed design and layout study, and would also demand an analysis of the likely yield/cost implications based on statistical data for the fabrication process utilised. It seems likely to the authors that the development of distinct MMIC chips rather than a single integrated chip is the most prudent option, at least for a first design iteration.

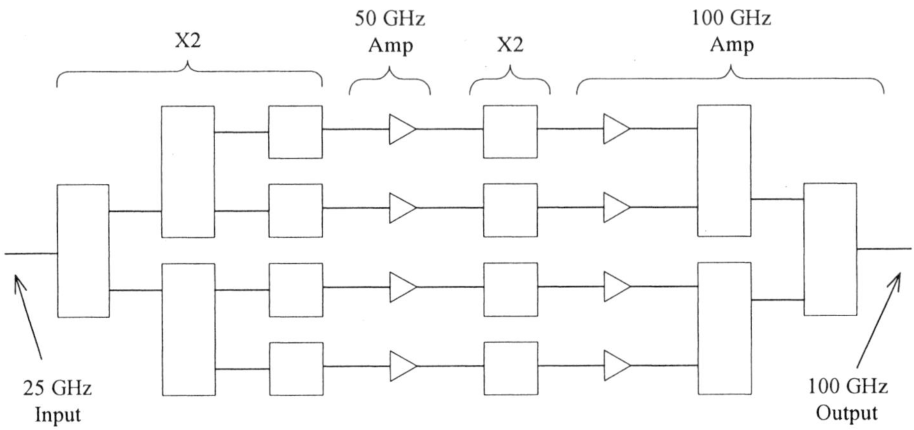

Figure 7-61. Second alternative X2X2 chain configuration (Alt#2).

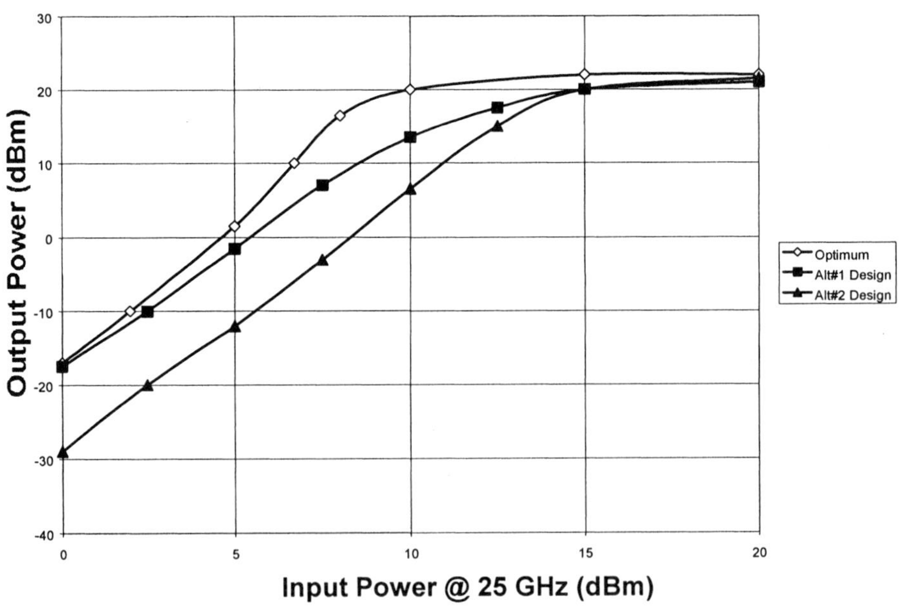

Figure 7-62. Comparison of 100 GHz output power versus 25 GHz input power characteristics for the optimum, Alt#1 and Alt#2 X2X2 chain configurations.

The outcome of this feasibility study was considered sufficiently successful that it is now being pursued to a fabrication phase in an ESA programme in conjunction with Farran Technology Ltd. and the Fraunhofer Institute[77].

6. MULTIPLIER CONCLUSIONS

In this chapter, a broad range of mm-wave FET frequency multiplier design topics have been discussed. A novel theoretical analysis of a single-ended FET multiplier has been demonstrated[72]. The more conventional analysis has been extended to allow for arbitrary bias and drive conditions. The results of this new analytical approach, in terms of the multiplying efficiency of a FET, were presented as a contour plot in the V_{GS} *vs* V_{AC} plane. This has yielded a very powerful visual summary of the multiplying capability of a FET and, at a glance, suggests the optimum bias and drive mix. It has also been shown through this analysis that a generic FET device may be capable of multiplier efficiency superior to that suggested by the more conventional analysis which does not allow for arbitrary double-sided clipping of the drain current waveform.

The theoretical analysis has been validated in a number of ways. Firstly, it yields results that are consistent with those predicted by the more conventional analysis over the range of bias/drive conditions for which that conventional approach is valid. Secondly, we have shown that this novel generic analysis yields results that are similar to those which are obtained from a purely numerical Fourier analysis of the current waveform of a FET based on the Angelov model[29]. Thirdly, the predictions were compared favourably with the measured performance of a frequency tripler and a doubler, both of which have been designed and evaluated during the course of this work.

Balanced multipliers are often used to address two shortcomings of high frequency multipliers, namely the difficulty of rejecting unwanted harmonics in some cases, and also the limited output power which can be obtained from a single device. A detailed analysis of balanced multipliers has been presented. This includes a discussion of some of the configurations that are most suited to high frequency use for various requirements. Some published balanced multiplier implementations were also reviewed. In the context of balanced multipliers, a novel flexible multiplier circuit was proposed. This offers the potential for electronic control of the multiplying factor of a balanced circuit with some additional rejection of all unwanted harmonics over and above that offered by each individual single-ended circuit.

Finally, an analysis of the design approach to high frequency multiplier chains was presented. This study examined the possibility of developing a monolithic multiplier/amplifier chain to generate 100 mW at 100 GHz, using a state of the art device. The various building blocks were discussed and the optimum configuration identified. The net outcome of this analysis was two-fold. Firstly, it was apparent that the generation of the target 100 mW was difficult, but realisable. Secondly, and equally importantly, a simple but very effective method has been demonstrated for taking a multiplier chain specification and, knowing the performance capability of the individual building blocks, deducing the optimum arrangement of building blocks to address the requirement. This work has evolved into a real wafer fabrication phase, where the results of this analysis can be validated[77].

It is certain that the future will see a continuing effort to develop frequency multipliers operating at higher frequencies and delivering more output power. This will involve developments similar to that studied in Section 5 of this chapter. Additionally, it is likely that more effort will be expended developing MMIC processes with high quality varactor compatibility. This will facilitate MMIC frequency multipliers with higher multiplication factors than those that are made possible by FETs alone.

Chapter 8

PRACTICAL MONOLITHIC TRANSCEIVER

1. INTRODUCTION

As outlined in Chapter 4, a key necessary step to enable the widespread use of monolithic technology in mm-wave systems, and thereby to facilitate the progression towards truly high volume mm-wave developments, is the realisation of monolithic transceivers at mm-wave frequencies. Subsequent to the mm-wave circuit developments described in Chapters 5 - 7, a fully monolithic transceiver MMIC was designed, laid out and fabricated. This design involved the integration of some of the individual building blocks from the earlier chapters into a single circuit layout. A novel custom package was developed to facilitate the testing of the resulting MMIC. In this chapter, the transceiver design is presented, and the layout discussed. Details of the package and assembly are outlined, and the measured performance is compared with expectation. The likely future of monolithic transceivers is also considered.

2. TRANSCEIVER TOPOLOGY

The transceiver topology adopted for this work, as discussed in Chapter 4, is shown in Fig. 8-1. The transmit chain consists of a power amplifier and an up-converting mixer. The receive arm contains a mixer and low noise amplifier. The LO for both arms is derived from a low frequency high quality input source, frequency multiplied (tripled in this case) and split by means of an active splitter.

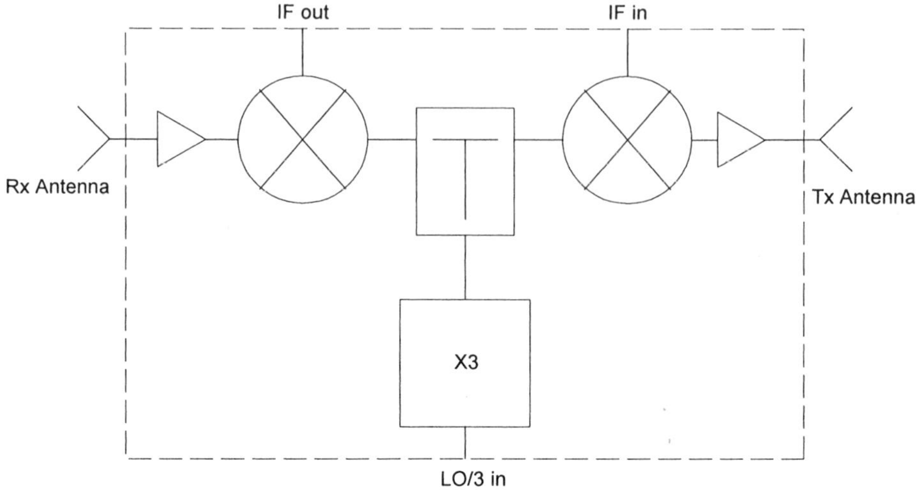

Figure 8-1. MMIC transceiver topology.

The intended application is the 57 GHz transceiver market. This market is ideally suited to short hop links and, due to the high intrinsic atmospheric attenuation in this band, does not present significant issues in terms of interference between nearby links. As a consequence, the band is loosely regulated in many countries and this makes it very attractive for a wide range of link requirements. This includes security or CCTV links, which can be easily installed and moved without the requirement for expensive cabling and infrastructure services.

The 57 GHz LNA and balanced diode mixer, described in Chapters 5 and 6 respectively, were used for the transceiver development. Also incorporated was the 57 GHz frequency tripler described in Chapter 7. A 57 GHz power amplifier, developed by a colleague of the authors in parallel with the LNA design[49], is also incorporated. This amplifier is similar in both schematic and layout to the 57 GHz LNA, and was fabricated on the same H40 process. The final element of the transceiver is the splitter – it is an active splitter consisting of a Wilkinson splitter in conjunction with a single stage of the power amplifier circuit in each arm. Some design adjustments were made to the building block circuits as a consequence of the absence of bond-wires on some of the internal connections. Specifically, this involved the internal connections between the mixers and the LNA and power amplifier, the mixers and the splitter, and between the frequency tripler and the splitter. The absence of inter-connecting bond-wires at these junctions demanded that the corresponding input – output matches be redesigned to compensate accordingly.

3. TRANSCEIVER LAYOUT

The transceiver layout is shown in Fig. 8-2. This is based very strongly on the layouts of the individual building block MMICs, with the interconnections being implemented as optimally as practical. The external RF connections are shown, and the individual building blocks locations in the layout are indicated. The same layout is again shown in Fig. 8-3 where the interconnects between the building blocks are indicated by the bold lines.

For ease of packaging, the DC bond pads are brought to the die edge. No effort was made to reduce the die area substantially. The primary purpose of this die layout was to assess the feasibility of such a complex multi-function single-chip design – die size optimisation should only be considered when an operating fully integrated design has been demonstrated. This approach was adopted to minimise any likelihood of undesirable coupling effects compromising the transceiver performance. The size of the transceiver die is 7.5 mm x 7.3 mm.

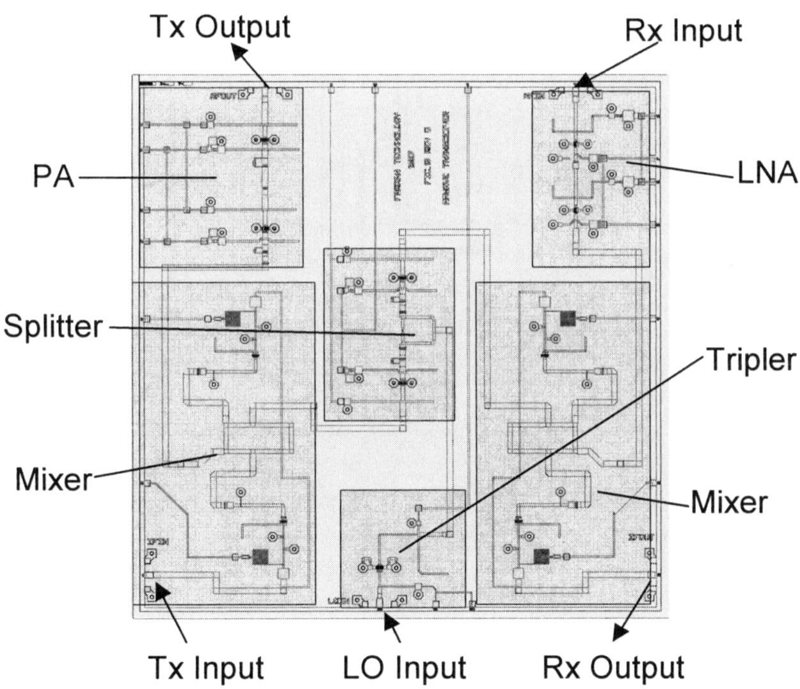

Figure 8-2. 57 GHz transceiver MMIC layout.

The transceiver chip was assembled into a custom metal housing to facilitate performance evaluation. The assembly drawing for this transceiver is shown in Fig. 8-4, and a photograph of a packaged transceiver MMIC in its housing is presented in Fig. 8-5. 50 Ω microstrip transmission lines on fused quartz can be seen between the MMIC and the coaxial connectors at the edges of the housing. V-type connectors are used for the mm-wave interfaces. SMA-f connectors are used for the lower frequency connections. DC filtercons are employed for the biasing inputs. Silver epoxy is used to attach the MMIC die and fused quartz sections to the gold-plated base of the aluminium housing. The Au bond-wires are 17 μm in diameter. The assembly effort associated with the packaging of this MMIC was entirely manual, and involved very intricate die attach and wire-bonding operations. As a consequence, only one die was assembled and tested. This die was selected at random as no 'known-good-die' evaluation technique was available.

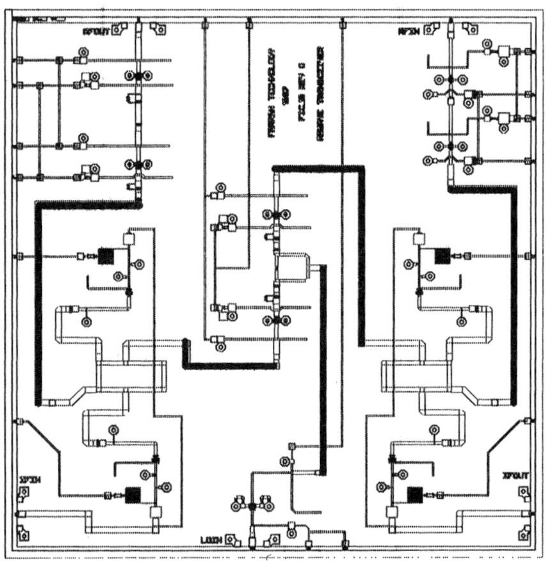

Figure 8-3. 57 GHz transceiver MMIC layout with interconnects between building blocks highlighted.

4. EXPECTED PERFORMANCE

The expected conversion performance of the fully integrated transceiver is summarised in Table 8-1. The up- and down-conversion characteristics are inferred from the individual building block measurements reported in Chapters 5 and 6, and in O'Ciardha et al[49] for the up-converting power amplifier.

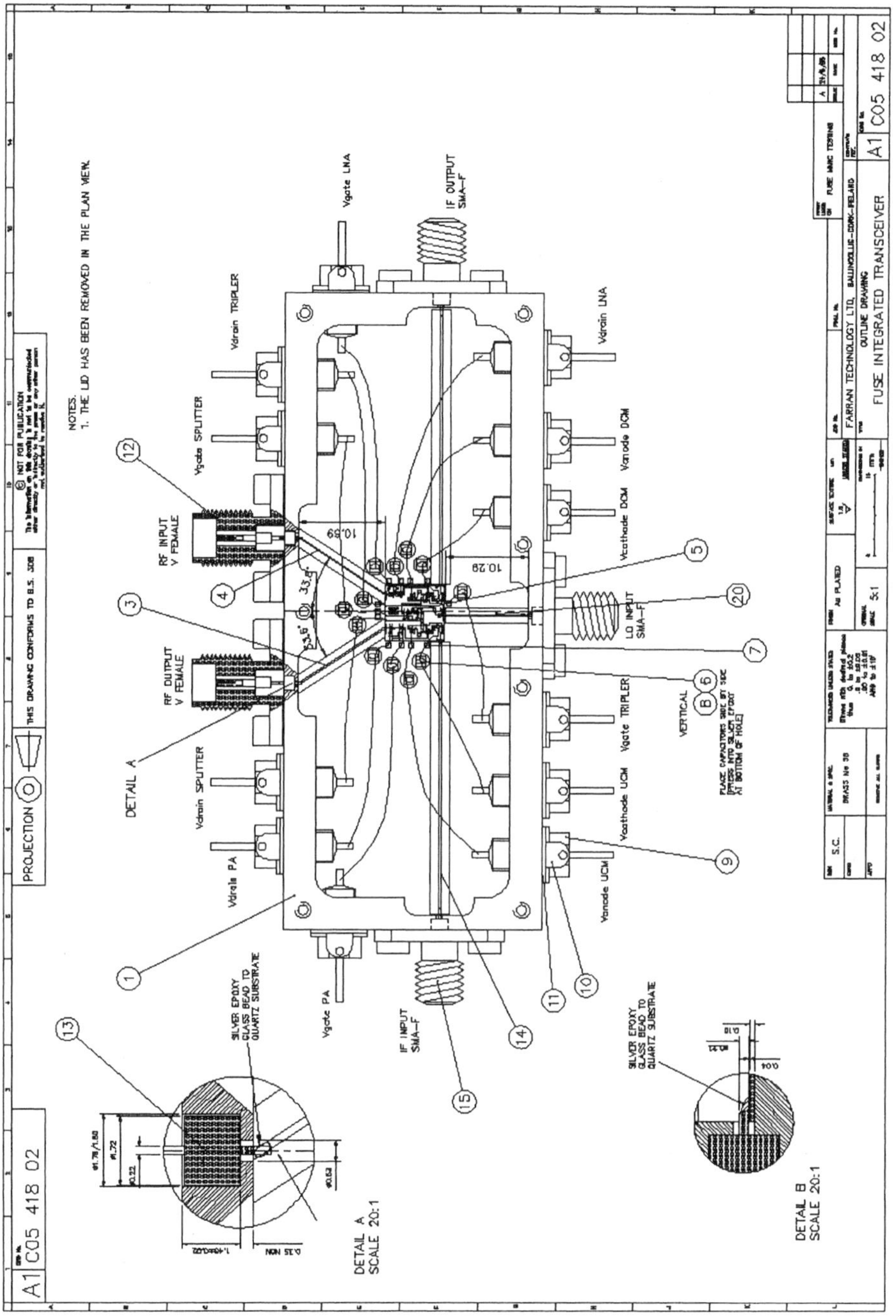

Figure 8-4. Assembly drawing for 57 GHz MMIC transceiver die.

Figure 8-5. Photograph of packaged transceiver MMIC die.

Table 8-1. Inferred transceiver conversion performance

Contribution to Conversion Performance	Up-Converter	Down-Converter
Mixer Loss (dB)	17	19
Amplifier Gain (dB)	8	14
Total Transceiver Loss(dB)	9	5

The expected performance for the down-converter is better than that for the up-converter primarily because of the higher gain associated with the LNA compared to the power amplifier.

5. MEASURED PERFORMANCE

The individual building blocks within the transceiver biased up correctly, and the packaged transceiver, as a whole, was functional in both up and down-converting modes of operation. The overall performance of the transceiver was not as good as expected, with a conversion loss of approximately 30 dB being achieved in both the up- and down-converting modes. Due to this measured performance, it was concluded that there would be no benefit in carrying out any further performance characterisation on the transceiver.

Given that no pre-screening of the die was possible, it is possible that a defect exists within one or more of the packaged transceiver's blocks; the fact that these blocks bias sensibly does not mean that their RF performance is guaranteed. As indicated in the yield analysis of Chapter 4, the large size of this transceiver die would suggest that the presence of at least one defect is certainly possible. For this approach to become viable with reasonable yield, die-size reduction would need to be pursued.

It is believed by the authors that the measured performance can be attributed to inadequate LO drive and conversion loss problems within the mixers. As indicated in Table 8-1, the measured conversion loss of the balanced diode mixer, when optimally pumped, was 17 dB for up-conversion, and 19 dB for down-conversion (see Chapter 6). Despite the fact that the measured transceiver up- and down-conversion losses are both significantly greater (by over 20 dB) than the expected losses shown in Table 8-1, it should be noted that those measured results are similar to each other (approximately 30 dB conversion loss for both up- and down-conversion modes of operation). This strongly suggests that a common defect may be affecting both the transmit and receive arms. This would be consistent with the output power capability of the active splitter being inadequate, resulting in low LO power being applied to the mixers– inadequate LO drive for the balanced diode mixers will of course degrade the conversion performance. Each arm of the active splitter contains a single-stage buffer-amplifier derived from the transceiver's 57 GHz power amplifier. An mentioned previously, both the active splitter and the 57 GHz power amplifier were developed by a colleague of the authors[49]. It was found that the measured compression characteristic of the 57 GHz power amplifier (tested independently from the transceiver) produced an unexpected result; the output 1dB compression point at 53 GHz was measured to be +9 dBm[49], but this degraded rapidly at higher frequencies and was of the order of 0 dBm in the LO frequency range. This strong frequency-sensitivity of the compression characteristic remains unexplained – discussions with the support engineers at the GMMT foundry[3] indicated that such behaviour had not been reported to them in the past. It is possible that attempting to use a transistor device close to its f_t may be a contributor to this effect, but the mechanism remains to be understood. However, taking account of the measured compression characteristic of the power amplifier at the LO frequency, it is almost certain that the output power capability of the active splitter is also limited and that the high conversion losses of both the up- and down-converters are caused by LO starvation of the balanced mixers. It should be possible to resolve this difficulty, but this could require the use of a pHEMT process with improved capability near 60 GHz.

It should also be pointed out that the new LNA design methodology, outlined in Chapter 5 in the context of a 30 - 40 GHz broadband design, lends itself very well to integration in a single-chip transceiver. As the measurements on this LNA indicate, the performance realised using this new approach is predictable and broadband, characteristics which are critical to successful integration with other transceiver elements.

6. CONCLUSIONS

The packaged transceiver has been found to be functional in both up and down-converting modes. Considering the complexity of the multi-function chip involved, and especially taking into account the high frequency range in the context of the fabrication process used, this is not an insignificant achievement.

The actual performance realised does not tally with that expected on the basis of the measured performance of the individual building blocks used in the integrated transceiver. It is believed by the authors that the active splitter design may require further optimisation to facilitate a higher LO drive for the two mixers. The die size, moreover, should be reduced to simplify the packaging complexity, and in particular to improve the circuit yield.

Nevertheless, it is clear that such a complex mm-wave transceiver can be realised. Performance improvements are clearly necessary to make this a viable option for mm-wave systems, but such improvements are realistic in the context of the high frequency capability of a typical fabrication process, increased LO drive to the individual mixers and die size optimization (for yield and cost). It is also likely that future trends in mm-wave transceiver development will make possible the integration of HBT and pHEMT technologies, thereby introducing the potential for a high-performance mm-wave oscillator on the same transceiver die. This increased level of integration will be a continuing trend in transceiver circuits, and the ultimate goal of a full system on a chip will come ever closer to reality.

Increased levels of integration are also driving a lot of research into alternative material systems, in particular SiGe. This compound semiconductor offers many of the advantages of GaAs based systems when used with Si in heterojunction type structures. In this way, Si based HEMT devices are made possible. Though these devices are not yet capable of delivering good mm-wave performance, the processes are continuously being improved and it is only a matter of time before mm-wave SiGe MMICs become a reality. The real "Holy Grail" here is the integration of front-end high frequency circuitry with low frequency or baseband CMOS components on a single Si substrate. The so-called "radio on a chip" development would furnish a solution with which GaAs simply could not compete[82, 83]. Of course, GaAs processes will continuously evolve also and will extend the frequency limits to which MMIC developments can aspire.

Chapter 9

GENERAL DISCUSSION AND FUTURE TRENDS

1. INTRODUCTION

In this chapter, the material described during the course of this work will be considered, in particular in the context of its place in the overall field of mm-wave integrated circuits. The original work will be specifically discussed in terms of its contribution to the field.

Likely future developments will be assessed, both in terms of possible extensions of some of the work presented here, and also in the context of potentially exciting areas of research in the general mm-wave integrated circuits field.

2. THIS WORK AND ITS PLACE IN THE GENERAL FIELD

The primary focus of the material presented in this work has been mm-wave transceiver developments in monolithic technology. This has been studied in the context of building blocks for two specific requirements. These are the emerging 57 GHz short-hop radio link and the 40 GHz MVDS applications.

The building blocks consist of amplifiers, mixers and frequency multipliers. All these components can be used in various combinations in the formation of receivers, transmitters or full transceivers. Individually, monolithic circuits have been reported with extremely high-frequency performance[84]. Such circuits typically have been reported using state of the art high frequency fabrication processes. All the circuits described in this work and summarised below have been fabricated using the GMMT H40 GaAs pHEMT process[3]. The process f_t is about 70 GHz and it is true to say that this process is representative of many volume manufacturing processes worldwide, but it certainly does not offer state of the art high frequency capability. It is in this context that the circuits outlined in this work must be assessed. The circuits' performance levels should be assessed in the context of the limited performance made possible by the H40 process at mm-wave frequencies. Original aspects of the work, either in terms of novel circuit design approaches, or new methods to analyse and understand circuit performance, will be emphasised in this section.

2.1 Amplifiers

The main performance results realised with MMIC amplifier designs described in Chapter 5, and in[85, 86], are summarised in Table 9-1. Two distinct low noise amplifiers have been developed. Initially, a two-stage 57 GHz circuit was designed and fabricated. Subsequently, a four-stage circuit was developed for 40 GHz applications.

Table 9-1. MMIC amplifier performance summary

Description	Gain (dB)	Noise Figure (dB)	Output P1dB (dBm)
57 GHz LNA	14.0	12.0	-1.0
40 GHz LNA	> 20 (26 – 44 GHz)	4.5 (35 – 40 GHz)	+11.0

The design approach adopted for the 57 GHz LNA was based on traditional methods. This consisted of designing narrow-band matching networks to optimise the circuit performance in the band of interest. No effort was made to make the circuit response broadband or to make the circuit performance tolerant of process and assembly variations. The performance realised with this design did not quite match that expected from simulation. Nevertheless, a very respectable gain (14 dB) was achieved near 57 GHz. To the authors' knowledge, this represents the highest frequency of operation reported for an amplifier fabricated on this H40 process.

An original design methodology was developed for the 40 GHz LNA. This approach yielded a high quality broadband circuit whose measured performance is consistent with simulation. The design is modular, with an excellent gain per unit area figure of merit. A simple, yet powerful and effective, bond wire compensation technique was developed and successfully implemented in this circuit.

2.2 Mixers

The key performance metrics realised with MMIC mixer designs described in Chapter 6 of this work are outlined in Table 9-2. The mixer circuits developed in this work were for the 57 GHz transceiver requirement. To this end, a mixer that is suitable for integration with other transceiver components in a multi-function die was desirable. Moreover, a circuit that is suitable for use as both an up- and a down-converter would be beneficial. It was found that a balanced Schottky diode mixer could meet these objectives. As no specific diode component is supported on the H40 process, the diode was implemented by using a pHEMT device with the source and drain shorted. The conversion performance realised with this balanced diode mixer was approximately 6 dB less than simulated. The reasons for this have been postulated in Chapter 6, but it remains true to say that, to the authors' knowledge, this mixer represents the highest frequency mixer developed on the H40 process.

Table 9-2. MMIC mixer performance summary

Description	Conversion Loss (dB)
57 GHz pHEMT Down-converter	10.0
57 GHz Diode Mixer as Down-converter	19.0
57 GHz Diode Mixer as Up-converter	17.0

A balanced pHEMT down-converting mixer has also been developed. This circuit offers the potential of superior performance to the diode mixer, but at the cost of a more complex IF hybrid requirement and distinct circuit designs for the up- and down-converters. The conversion performance realised with the fabricated mixer die in a package was again about 3 dB less than simulated, at about 10 dB. Nevertheless, such performance is reasonable considering the process limitations and the fact that this was a first design spin.

A novel aspect of both of the above balanced mixers was the incorporation of a reduced size ratrace element. This makes possible reduced die area for the mixer without compromising the circuit performance over a slightly narrower band. In fact the reduced size makes possible lower ohmic losses in this hybrid.

In the context of improving the fundamental understanding of the operating principles behind FET mixer circuits, an interesting approach has been developed and is presented in Chapter 6. This analysis outlines how the optimum bias point is found for a given topology of FET mixer. The critical FET characteristics for various topologies are identified and the selection of the best biasing conditions justified in each case. This basic approach supplements the detailed circuit simulations which are more accurate but provide less insight into the circuit design tradeoffs.

2.3 Frequency Multipliers

The performance realised with the two MMIC frequency multiplier designs described in Chapter 7 of this work is presented in Table 9-3.

Table 9-3. MMIC frequency multiplier performance summary

Description	Conversion Loss (dB)
56 GHz Tripler	14.1
40 GHz Doubler	-1.5 (i.e. conversion gain)

Two frequency multiplying circuits have been developed. A frequency tripler was designed for the 57 GHz transceiver application, and a 40 GHz doubler was developed for the MVDS market.

The tripler circuit consists of a single-ended stage with reactive networks at both input and output. The pHEMT device is biased near pinchoff to ensure good current waveform clipping and harmonic generation. The conversion performance realised with the fabricated circuit was about 3 dB less than simulated, which was an encouraging result given the frequency of operation. The 3 dB discrepancy is mainly attributed to limited model accuracy in the frequency range of interest.

The frequency doubler design also incorporated an output amplifying stage. This approach enhances the output power capability of the circuit, and the bandwidth characteristics of the amplifier also facilitate the unwanted harmonic rejection properties of the doubler as a whole. The performance realised with this fabricated doubler compared very well with expectation and justifies the adoption of the approach incorporating the amplifying stage.

3. FUTURE DEVELOPMENTS

In this section, likely future areas of research and development are discussed. In the context of expansions on the work reported here, specific follow-on activities are outlined. In the broader context of the mm-wave monolithic circuit field, the authors' opinions about future directions in terms of technology, process and circuit development are presented. Of course the future is difficult to predict, but there are specific areas of research and development in which there are likely to be exciting breakthroughs over the next five to ten years.

3.1 Based on This Work

Further development of MMIC designs for the 57 GHz radio-link market hinges entirely on a suitable commercially available foundry fabrication process being identified. It is clear from the work carried out here that the GMMT H40 process is not capable of providing adequate performance in V-band, in particular in the case of active circuits. A process with an f_t in the region of at least 90 GHz would be recommended to make the realisation of high performance 57 GHz designs with reasonable yield possible.

In terms of the circuits themselves, it is clear that a balanced diode mixer offers greater flexibility in terms of suitability for transceiver integration, but the successful design of a monolithic diode mixer circuit absolutely requires an optimised Schottky diode device that is fully characterised and modelled by the foundry.

The amplifier approach adopted for the 4-stage 40 GHz LNA could be ported to an alternative process and centred at 57 GHz with relatively little difficulty, provided the design methodology outlined in Chapter 5 is followed.

The circuits developed for the 40 GHz market do not require any further design effort in the context of the H40 process – however, if it were deemed desirable to have these circuits fabricated on an alternative process, such a redesign effort could be carried out with low risk. Such an undertaking could possibly be driven in the future by market demands for even lower receiver noise figure and such a requirement might necessitate the use of a higher performance fabrication process.

3.2 General MMIC Field

A number of likely trends in terms of process development have been mentioned in this work.

Firstly, in order to facilitate enhanced integration of mm-wave functions, there is an ever-growing need for high quality MMIC oscillators. These can be implemented as either low-frequency sources driving mm-wave frequency multipliers, or alternatively as direct mm-wave oscillators. The latter clearly is the preferred approach in terms of reduced circuit complexity. Regardless of the approach, for full integration, high performance GaAs based oscillators are required. Conventional FET devices are not suitable for oscillator development due to their poor $1/f$ noise characteristics. Bipolar transistors are generally used in the design of lower frequency oscillators. With the emergence of HBT devices, high quality oscillators are becoming a reality on GaAs. HBT devices are also eminently suited to high power applications including power amplifiers and output stages of frequency multipliers where appropriate. In the context of transceiver integration, the difficulty with HBT devices is that they do not generally support good low noise performance. In other words, even if a high quality HBT based mm-wave oscillator and power amplifier were to be developed, the likelihood is that FET based circuits would still be required for the LNA and probably the mixing functions and hence a mix of technologies would be involved in the final transceiver implementation – thereby negating the efforts to integrate all functions into a single die! The development which will address this issue is the integration on a single substrate of HBT and pHEMT devices. Clearly, there are significant technical obstacles in the way of achieving this, but there is no doubt about the clear benefits that such a combined process will offer. This will facilitate the development of mm-wave transceivers with all functions using the optimum device flavour to optimise performance. In this context, significant progress has been published[87, 88], where HEMT-HBT fabrication capability has been reported on both GaAs and InP substrate material. These processes involve selective re-growth (using MBE) of HEMT islands on patterned and etched HBT material. Such processes, though expensive and non-standard, offer the potential for increased levels of integration of circuit functions in the future.

The next step in terms of integration is to include the front-end analog components with the baseband digital circuits. This can be considered in two ways.

Firstly, there is the possibility of implementing the digital circuits on GaAs. Digital GaAs circuits have been reported for many years[7], and especially high-speed performance is possible using this technology. It is conceivable that baseband circuits could be developed on GaAs to integrate with front-end analog circuits. However, not alone would this be extremely uncompetitive in terms of cost compared with a conventional CMOS equivalent on silicon, but the GaAs digital circuit would also suffer from a relatively high current drain. The cause of this increased current drain is the non-zero MESFET gate current. As was mentioned in Chapter 3, the development of MOSFET devices on GaAs is of interest in this context[28].

However, even if acceptable GaAs MOSFET devices become a reality, the total implementation on GaAs is still probably not viable due to cost. The alternative view is to consider whether or not the total requirement can be implemented on silicon instead. The obvious issue here is how to realise good performance at high frequencies using silicon technology. Much research is ongoing in this area. SiGe devices are offering the most promising results. These are essentially heterojunction devices fabricated on a silicon substrate. They offer the enhanced performance delivered by heterojunction structures, but are compatible, at least to a significant degree, with conventional silicon fabrication technology. Besides the device technology, the other great issue to be overcome to make silicon a viable substrate material for mm-wave developments is the difficulty of realising high quality low loss passive structures. For instance, spiral inductors on silicon tend to be very lossy due to the low resistivity (relative to GaAs) of typical CMOS substrate material. Generally, the parasitic substrate effects tend to degrade high frequency circuit performance to a great degree. As a consequence, research efforts are focussing on ways to separate the passive structures from the main silicon bulk, primarily by incorporating additional dielectric layers in between. This helps to reduce the parasitic effects, but potentially adds to the fabrication costs and complexity. It is certain that these developments will result in silicon processes that support high frequency developments integrated with baseband circuits, thereby making possible full radios on a single chip. There is a strong body of opinion[89] which believes, for example, that the automotive collision avoidance developments at 77 GHz will only become a true high volume reality when they can be fabricated fully on silicon processes. This is probably unrealistic, and certainly such radar products are already being sold in luxury class cars based on a mix of silicon, monolithic GaAs and Gunn diode technologies. It is the authors' opinion that further integration of these units using existing GaAs technology will be sufficient to enable the more widespread incorporation of these radar products in all cars. However, it is equally true that provided Si/SiGe developments reach a level of maturity where 77 GHz transceivers can be manufactured with reasonable yields, then the cost structure will be such that GaAs based equivalent units will be uncompetitive.

However, there is also no doubt that in terms of high frequency performance, GaAs based technology will always have an advantage over silicon. In particular, as the frequency spectrum becomes increasingly over-crowded, there is a demand for systems operating at higher and higher frequencies, and this trend is certainly going to continue. As silicon technology makes advances towards system on a chip solutions at frequencies up to and including low mm-wave frequencies, the new and yet to emerge system requirements will begin to enter the higher mm-wave frequency range where only GaAs (and other III-V materials) can be used. On this basis, one can predict major developments in both the silicon/SiGe and the higher mm-wave frequency range GaAs process technology fields in the years to come.

Other areas of ongoing research involve new material systems. InP offers the potential for superior performance at high frequencies, but due to its low barrier height, its use is limited as a result of leakage and breakdown effects. This is particularly the case for high power applications. The new materials include GaN, which is of particular interest for high power and high temperature applications due to its large barrier height. It should be noted that GaN is also of interest for opto-electronic applications in the blue/UV spectral region, and this dual use results in a significant flow of R&D funding for process and material development in this area. The University of Florida[90] and Kansas State University[91] are examples of research groups working on GaN developments. The key issue with GaN is that it cannot (as yet) be grown in bulk, and can only be grown on top of another material, often SiC. It can be assumed that this will be solved in time, and then GaN will become a very attractive semiconductor material for a whole host of demanding volume applications. An interesting review article on this topic has been published[92].

As has been made clear throughout this work, a successful mm-wave MMIC design absolutely requires that the die packaging approach be built into the design. In this work, the MMICs developed were all packaged using bond wires and an effective bond wire compensation technique has been demonstrated. However, it is clear that the use of such bond wires becomes less and less attractive at higher frequencies. Flip chip packaging techniques are becoming increasingly attractive, particularly at high frequencies due to the reduced parasitic effects. However, this technology is less mature than wire bonding and extensive research is being conducted in this field[93].

An example of a company carrying out research in this area is Endwave[94]. The motivation behind Endwave's developments is that much of the GaAs area associated with many complex MMIC designs is implementing bias circuitry and other passive structures which could be equally well implemented in cheaper technology provided that the total integrated solution is kept small. Endwave have reported significant cost savings by reducing the MMICs to the bare essentials (mainly the active elements) and mounting these using patented flip-chip techniques on a single ceramic substrate on which the other necessary structures are also patterned. The overall result is compact, and according the Endwave's claims, cheap to manufacture. As the GaAs dice are much smaller, the yield from a given wafer is also expected to improve.

Other work is being carried out developing true surface-mount packaging solutions for MMICs. One example, a result of product development at HD Communications Corp.[95], is shown in Fig. 9-1. The VBGA package is manufactured using semiconductor fabrication techniques, thereby enabling significant cost savings. The base material for the package is Via Plane™, a single layer ceramic interconnect substrate with tungsten-copper vias. This ceramic substrate material is 99.6% alumina. Good high-frequency performance is facilitated by low inductance vias and a tight control over patterned line impedances using thin-film processing techniques. The vias are used to provide good grounds, both RF and thermal. Cu-Ag balls form the I/O interface to the PC board (e.g. RT-Duroid). A ceramic or plastic lid can be used to encapsulate the MMIC. Excellent flatness characteristics have been reported for the VBGA substrate[95], which is a key requirement to make possible high assembly yields for the thin GaAs mm-wave MMIC die.

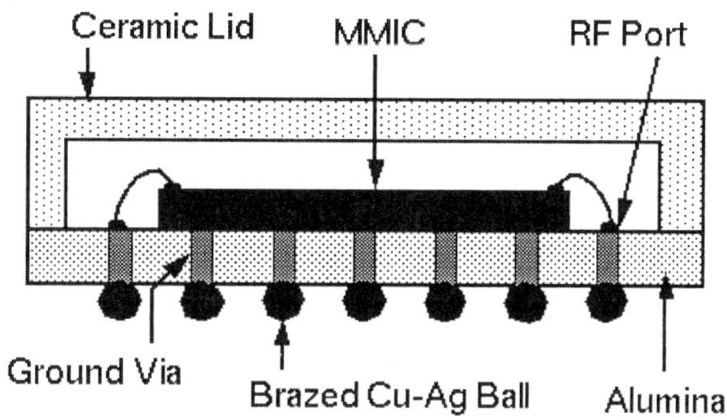

Figure 9-1. VBGA package cross-section (source HD Communications Corp.[95]).

In terms of circuit design, there is a common belief that few opportunities exist for original circuit development. The popular view is that circuit design consists of taking existing design approaches and adapting them to suit particular applications. However, it is the authors' belief that the opportunity still exists for novel circuit design approaches to be developed.

One way of considering this would be as follows. As more and more integration becomes possible with advances in MMIC technology and associated process/product yield improvements, the interface between building blocks should become an area of research and optimisation. There may be considerable scope to reduce circuit complexity by combining functions at the output of one block and the input of the next block into a single less complex network. A simple example should help to illustrate this in principle. Consider an LNA driving a mixer. As standalone blocks, the LNA has an output matching network and biasing injection scheme. Similarly, the mixer will have some form of input matching and biasing arrangement. When integrated, these networks could be combined, certainly in terms of the matching and, with prudent design, possibly in terms of the biasing as well. In fact, it is almost certain that by combining the matching networks in this way, a more effective overall broadband match could be realised.

References

1. T. Jackson, "A Novel System for Providing Broadband Multimedia Services to the Home - The Mesh Network", *Microwaves & RF 98 Conference Proceedings,* Wembley, London, pp. 86-91, Sept. 1998.
2. H. H. Meinel, "The Market for Short-Haul Line-of Sight Millimeter Wave Transmission Links", *IEEE MTT-S International Microwave Symposium Digest,* San Francisco, 1996, pp. 487-489.
3. http://www.bookham.com/rf/about-caswell.htm
4. S. M. Sze, "*Physics of Semiconductor Devices*", 2nd Edition, John Wiley & Sons, Inc., New York, 1981, ISBN 0-471-05661-8.
5. R. S. Muller, T. I. Kamins, "*Device Electronics for Integrated Circuits*", 2nd Edition, John Wiley & Sons, Inc., New York, 1986, ISBN 0471887587.
6. S. A. Maas, "*Microwave Mixers*", Artech House, Inc., Dedham, MA, 1986, ISBN 0-89006-171-8.
7. S. I. Long, S. E. Butner, "*Gallium Arsenide Digital Integrated Circuit Design*", McGraw-Hill Publishing Company, 1990, ISBN 0070386889.
8. D. K. Ferry, "*Gallium Arsenide Technology*", Howard W. Sams & Co., Inc., Indianapolis, 1985, ISBN 0-672-22375-9.
9. G. D. Vendelin, A. M. Pavio, U. L. Rohde, "*Microwave Circuit Design Using Linear and Nonlinear Techniques*", John Wiley & Sons, Inc., New York, 1990, ISBN 0-471-60276-0.
10. D. Pavlidis, "HBT vs. PHEMT vs. MESFET: What's best and why", *International Conference on GaAs Manufacturing Technology (MANTECH),* Vancouver, BC, Canada, May 1999, (http://www.gaasmantech.org/digest/1999/42.html).
11. D. Greenberg, S. Sweeney, G. Freeman, D. Ahlgren, "Low-Noise Performance Near BVCEO in a 200 GHz SiGe Technology at Different Collector Design Points", *IEEE*

MTT-S International Microwave Symposium Digest, Philadelphia, PA, 2003, pp. 113–116.

12. R. S. Pengelly, "*Microwave Field Effect Transistors - Theory, Design and Applications*", Second Edition, Research Studies Press Ltd., 1986, ISBN 0-86380-040-8.
13. J. L. Hudgins, "Wide and Narrow Bandgap Semiconductors for Power Electronics: A New Valuation", *Journal of Electronic Materials*, Vol. 32, No. 6, pp. 471-477, Junc 2003.
14. S. M. Sze, "*High Speed Semiconductor Devices*", John Wiley & Sons, Inc., New York, 1990, ISBN 0-471-62307-5.
15. F. Ali, I. Bahl, A. Gupta, "*Microwave and Millimeter-Wave Heterostructure Transistors and their Applications*", Artech House, Inc., Norwood, MA, 1989, ISBN 0-89006-372-9.
16. M. S. Tyagi, "*Introduction to Semiconductor Materials and Devices*", John Wiley & Sons, Inc., New York, 1991, IBSN 0-471-60560-3.
17. M. Shur, "*Introduction to Electronic Devices*", John Wiley & Sons, Inc., New York, 1996, ISBN 0-471-10348-9.
18. M. Shur, "*GaAs Devices and Circuits*", Plenum Press, New York, 1987, ISBN 0-306-42192-5.
19. Picoprobe by GGB Industries, Inc., http://www.ggb.com/index.html.
20. Kopin Corporation, http://www.kopin.com.
21. P. Win, Y. Druelle, A. Cappy, Y. Cordier, J. Favre, C. Bouillet, "Metamorphic $In_{0.3}Ga_{0.7}As/In_{0.29}Al_{0.71}As$ layer on GaAs: A New Structure for High Performance High Electron Mobility Transistor Realization", *Appl. Phys. Lett.*, Vol. 61, p 922, 1992.
22. J. J. Liou, "*Principles and Analysis of AlGaAs/GaAs Heterojunction Bipolar Transistors*", Artech House, Boston, 1996, ISBN 0-89006-587-X.
23. S. M. Sze, "*Modern Semiconductor Device Physics*", John Wiley & Sons, Inc., New York, 1998, ISBN 0-471-15237-4.
24. H. A. Bethe, "Theory of the Boundary Layer of Crystal Rectifiers", *MIT Radiation Lab. Tech. Report 43-12*, 1942.
25. W. Schottky, "Halbleitertheorie der Sperrschicht", *Naturwissenschaften*, 26, 843, 1938.
26. C. R. Crowell and S. M. Sze, "Current Transport in Metal-Semiconductor Barriers", *Solid State Electron*, 9, pp. 1035-1047, 1966.
27. http://eesof.tm.agilent.com/products/adsoview.html
28. Y. C. Wang, M. Hong, J. M. Kuo, J. P. Mannaerts, J. Kwo, H. S. Tsai, J. J. Krajewski, Y. K. Chen, A. Y. Cho, "Demonstration of Submicron Depletion-Mode GaAs MOSFET's with Negligible Drain Current Drift and Hysteresis", *IEEE Electron Device Letters*, Vol. 20, No. 9, pp. 457-459, Sept. 1999.
29. I. Angelov, H. Zirath, N. Rorsman, "A New Empirical Nonlinear Model for HEMT and MESFET Devices", *IEEE Trans. Microwave Theory Tech.*, Vol. MTT-40, No. 12, pp. 2258-2266, Dec. 1992.
30. N.I. Cameron, M.R.S. Taylor, H. McLelland, M. Holland, I.G. Thayne, K. Elgaid, S.P. Beaumont "A High Performance, High Yield, Dry-Etched, Pseudomorphic HEMT for W-

Band Use", *IEEE MTT-S International Microwave Symposium Digest*, Orlando, 1995, pp. 435-438.

31. O. Ishikawa, K. Nishii, T. Matsuno, C. Azuma, Y. Ikeda, S. Nanbu, K. Inoue, "Low-Noise InGaAs HEMT Using the New Off-Set Recess Gate Process", *IEEE MTT-S International Microwave Symposium Digest*, Long Beach, 1989, pp. 979-982.
32. K.H.G. Duh, P.C. Chao, P. Ho, M.Y. Kao, P.M. Smith, J.M. Ballingall, A.A. Jabra, "High-Performance InP-based Millimeter-wave Low-Noise Amplifiers", *IEEE MTT-S International Microwave Symposium Digest*, Long Beach, 1989, pp. 805-808.
33. C.A. Liechti, "Heterostructure Microwave Transistors for Ultra-High Speed Electronics", *European Microwave Conference Proceedings*, Stockholm, 1988, pp. 92-100.
34. H. Happy, S. Bollaert, K. Foure, A. Cappy, "Numerical Analysis of Device Performance of Metamorphic $In_yAl_{1-y}As/In_xGa_{1-x}As$ ($0.3 \leq x \leq 0.6$) HEMT's on GaAs Substrate", *IEEE Trans. Microwave Theory Tech.*, Vol. MTT-45, No. 10, pp. 2089-2095, Oct. 1998.
35. http://www.gaas.net/news/Pico_1097.html
36. D.A. Divekar, "*FET Modeling for Circuit Simulation*", Kluwer Academic Publishers, 1988, ISBN 0898382645.
37. W. Curtice, "A MESFET Model for use in the Design of GaAs Integrated Circuits", *IEEE Trans. Microwave Theory Tech.*, Vol. MTT-28, No. 5, pp. 448-455, May 1980.
38. Materka, T. Kacprzak, "Computer calculation of Large-Signal GaAs FET Amplifier Characteristics", *IEEE Trans. Microwave Theory Tech.*, Vol. MTT-33, No. 2, pp. 129-134, Feb. 1985.
39. H. Statz, P. Newman, "GaAs FET Device and Circuit Simulation in SPICE", *IEEE Trans. Electron Devices*, ED-34, 1987, pp. 160-166.
40. H. Kroemer, "Theory of a Wide-Gap Emitter for Transistors", *Proc. IRE*, Vol. 45, No. 11, pp. 1535-1537, Nov. 1957.
41. B. Lin, "InGaP HBTs Offer Enhanced Reliability", *Applied Microwave and Wireless*, Noble Publishing, Dec. 2000, pp. 115-116.
42. B. Lin, L. Wang, "InGaP emitters make HBTs more reliable", *GaAs Manufacturing*, May 2001 (http://www.compoundsemiconductor.net/articles/magazine/7/5/70/1).
43. http://www.cen.uiuc.edu/~gstill/research.html
44. J-G Kim, D-H Baek, S-H Jeon, J-W Park, S. Hong, "A 60 GHz InGaP/GaAs Push-Push MMIC VCO", *IEEE MTT-S International Microwave Symposium Digest*, Philadelphia, PA, 2003, pp. 885-888.
45. M. Wetzel, R.A. Johnson, P.F. Chen, B. Xavier, P.R. de la Houssaye, P. M. Asbeck, J. Lagnado, "Prospects of CMOS Power Amplifiers on Thin Film SOS for Wireless Applications", *First Annual UCSD Conference on Wireless Communications*, San Diego, CA, Mar. 1998 (http://sigma.ucsd.edu/~mwetzel/).
46. J. Yue, J. Kriz, "SOI CMOS Technology for RF System-On-Chip Applications", *Microwave Journal*, Horizon House Publications, Jan. 2002.
47. L. P. B. Katehi, G. M. Rebeiz, and C. T.-C. Nguyen, "MEMS and Si-micromachined components for low-power, high-frequency communications systems (invited)," *IEEE MTT-S International Microwave Symposium Digest*, Baltimore, June 1998, pp. 331-333.

48. http://www.ums-gaas.com/
49. E. O'Ciardha, P. Duffy, D. Vizard, B. Lyons, M. Fehilly, S. Lidholm, "MMIC Circuit Developments for Millimetre-Wave Applications", *Microwaves & RF Conference Proceedings*, Wembley, London, Sept. 1998.
50. K.H.G.Duh, P.-C. Chao, P.M. Smith, L.F. Lester, B.R. Lee, J.M. Ballingall, M.-Y. Kao, "High-Performance Ka-Band and V-Band HEMT Low-Noise Amplifiers", *IEEE Trans. Microwave Theory Tech.*, Vol. MTT-36, No. 12, pp. 1598-1603, Dec. 1988.
51. G. Gonzalez, "*Microwave Transistor Amplifier Analysis and Design*", Prentice-Hall, Englewood Cliffs, N.J., 1996, ISBN 0132543354.
52. E. O'Ciardha, P. Duffy, B. Lyons, M. Fehilly, S. Lidholm, "Fully Integrated Mmwave Radio Transceiver", *ESA Workshop on Mm-wave Technology Applications*, Helsinki, pp. 196-201, May 1998.
53. E. O'Ciardha, P. Duffy, D. Vizard, B. Lyons, M. Fehilly, S. Lidholm, "MMIC Circuit Developments for Millimetre-Wave Applications", *European Microwave Conference Proceedings*, Amsterdam, Oct. 1998, pp 58-63.
54. B. N. Lyons, E. O'Ciardha, P. Duffy, M. Fehilly, S. Cremin. "Design and Evaluation of a 60 GHz MMIC", *IEE Colloquium on MM-Wave Circuits and Technology for Commercial Applications*, London, March 1999, pp. 4/1-4/6.
55. H.A. Haus, R.B. Adler, "*Circuit Theory of Linear Noisy Networks*", MIT Press, Cambridge, MA, 1959, ISBN 0262080079.
56. http://www.rogers-corp.com/mwu/pdf/5000data.pdf
57. W.M. Kelly, G.T. Wrixon, "Conversion Losses in Schottky-Barrier Diode Mixers in the Submillimeter Region", *IEEE Trans. Microwave Theory Tech.*, Vol. MTT-27, No. 7, p. 665-672, July 1979.
58. M. Schefer, U. Lott, H. Benedickter, Hp. Meier, W. Patrick, and W. Bächtold, "Active Monolithically Integrated Coplanar V-Band Mixer", *IEEE MTT-S International Microwave Symposium Digest*, 1997, Denver, pp. 1043-1046.
59. I. D. Robertson, "MMIC Design", *Institution of Electrical Engineers*, 1995, ISBN 0852968 167.
60. J.P. Viaud, "Modular Millimetre-wave transceiver design for today's digital radio links", *Microwave Engineering Europe*, Feb 99.
61. J. Mizoe, T. Matsumura, K. Unosawa, Y. Akiba, K. Nagai, H. Sato, T. Saryo, T. Inoue, "A V-Band GaAs MMIC Chip Set on a Highly Reliable Wsi/Au Refractory Gate Process", *IEEE MTT-S International Microwave Symposium Digest*, 1997, Denver, pp. 247-250.
62. T.J. Ellis, "A Planar Circuit Design for High Order Sub-Harmonic Mixers", *IEEE MTT-S International Microwave Symposium Digest*, 1997, Denver, pp. 1039-1042.
63. A. Dearn (GMMT), Private communication, 1997.
64. D. I. Kim, G.-S. Yang, "Design of New Hybrid Ring Directional Coupler Using $\lambda/8$ or $\lambda/6$ Sections", *IEEE Trans. Microwave Theory Tech.*, Vol. MTT-39, No. 10, pp. 1779-1784, Oct. 1991.

65. Gopinath, J.B. Rankin, "Single-gate MESFET frequency doublers", *IEEE Trans. Microwave Theory Tech.*, Vol. MTT-30, No. 6, pp. 869-875, June 1982.
66. S.A. Maas, "*Nonlinear Microwave Circuits*", Artech House, 2003, ISBN 1580534848.
67. H. Zirath, I. Angelov, N. Rorsman, C. Karlsson, E. Kollberg, "A Balanced W-Band HFET Doubler", *European Microwave Conference Proceedings*, Madrid, 1993, pp. 837-839.
68. J.S. Augusto, M.J. Rosario, J.C. Vaz, J.C. Freire, "Optimal Design of MESFET Frequency Multipliers", *European Microwave Conference Proceedings*, Madrid, 1993, pp. 402-404.
69. R. Gilmore, "Octave-Bandwidth Microwave FET Doubler", *Electronics Letters*, Vol. 21, No. 12, pp. 532-533, 6th June 1985.
70. T.Hirota, H. Ogawa, "Uniplanar Monolithic Frequency Doublers", *IEEE Trans. Microwave Theory Tech.*, Vol. MTT-37, No. 8, pp. 1249-1254, Aug. 1989.
71. H. Fudem, E.C. Niehenke, "Novel Millimeter Wave Active MMIC Triplers", *IEEE MTT-S International Microwave Symposium Digest*, Baltimore, 1998, pp. 387-390.
72. E. O'Ciardha, S. Lidholm, B. Lyons, "Generic Device Frequency Multiplier Analysis – A Unified Approach", *IEEE Trans. Microwave Theory Tech.*, Vol. MTT-48, No. 7, pp. 1134-1141, July 2000.
73. K. Yhland, "FET Mixers and Nonlinear FET Models for Intermodulation Analysis", *Technical Report No. 287L*, Chalmers University of Technology, 1998.
74. I. Angelov, "Chalmers Nonlinear HEMT and MESFET Model - extraction procedure: Part A, Extraction of the current part of the model", *Department of Microwave Technology Report No. 25*, Chalmers University of Technology, July 1996.
75. E. O'Ciardha, J. Lynch, B. Lyons, S. Cremin, S. Lidholm, "Practical mm-Wave pHEMT Frequency Multiplier", *International Journal of Infrared and Millimeter Waves*, Vol. 23, No. 6, June, 2002.
76. E. O'Ciardha, B.N. Lyons. "100 GHz MMIC Power Source Development Study", *International Journal of Infrared and Millimeter Waves*, Vol. 21, No. 11, November 2000.
77. "100 GHz Monolithic Multipliers and Power Amplifier", *Farran Technology Ltd. / ESA Contract 15820/01/NL/FM(SC).*
78. P.-P. Huang, T.-W. Huang, H. Wang, E. W. Lin, Y. Shu, G. S. Dow, R. Lai, M. Biedenbender, J. H. Elliott, "A 94 GHz 0.35W Power Amplifier Module", *IEEE Trans. Microwave Theory Tech.*, Vol. MTT-45, No. 12, pp. 2418-2423, Dec. 1997.
79. F. Brauchler, J. Papapolymerou, J. East, L. Katehi, "W Band Monolithic Multipliers", *IEEE MTT-S International Microwave Symposium Digest*, Denver, 1997, pp. 1225-1228.
80. P. Stenger, M. Sarantos, E. Niehenke, H. Fudem, C. Schwerdt, F. Kuss, D. Strack, G. Hall, J. Masti, "A Miniature MMIC One Watt W-Band Solid State Transmitter", *IEEE MTT-S International Microwave Symposium Digest*, Denver, 1997.
81. E. O'Ciardha, S. Lidholm, "Novel MMIC Frequency Multiplier Configuration", *International Journal of Infrared and Millimeter Waves*, Vol. 21, No. 7, July 2000.

82. J. Ouellette, "Radio on a Chip Expedites Wireless Web", *The Industrial Physicist*, pp. 9-12, Aug. 2000.
83. J. Lyman, "'Radio on a Chip' Debuts", May 2002: http://www.newsfactor.com/perl/story/17990.html
84. R. Lai, H. Wang, Y. C. Chen, T. Block, P. H. Liu, D. C. Streit, D. Tran, P. Siegel, M. Barsky, W. Jones, T. Gaier, "D-band MMIC LNAs with 12 dB gain at 155 GHz fabricated on a high yield InP HEMT MMIC production process", *International Conference on Indium Phosphide and Related Materials*, New York, 1997, pp. 241 –244.
85. E. O'Ciardha, J. Lynch, B. Lyons, P. Duffy, S. Cremin, S. Lidholm, "Low Noise mm-Wave Amplifier Design Methodology", *International Journal of Infrared and Millimeterwaves*, Vol. 22, No. 12, Dec. 2001.
86. E. O'Ciardha, J. Lynch, B. Lyons, P. Duffy, S. Cremin, S. Lidholm, "Novel Design Approach for mm-Wave Low Noise Amplifier", *Irish Signals and Systems Conference Proceedings*, University College Cork, June 2002, pp. 161-165.
87. K. W. Kobayashi, A. K. Oki, D. K. Umemoto, T. R. Block, "A Novel Monolithic HEMT-HBT Ka-band VCO-Mixer Design", *IEEE MTT-S International Microwave Symposium Digest*, 1997, Denver, pp. 83-86.
88. K. Wang, R. Lai, L. Tran, J. Cowles, Y. C. Chen, E. W. Lin, H. H. Liao, M. K. Ke, T. Block, H. C. Yen, "A Single-Chip 94-GHz Frequency Source Using InP-Based HEMT-HBT Integration Technology", *IEEE MTT-S International Microwave Symposium Digest*, Baltimore, June 1998, pp. 219-222.
89. P. Greiling, "HRL Technology Overview", *Presented at the National Microelectronics Research Centre*, Cork, Ireland, Sept. 1998.
90. University of Florida: http://www.mse.ufl.edu/~spear/recent_papers/
91. KSU: http://www.phys.ksu.edu/area/GaNgroup/index.html
92. U.K. Mishra, L.F. Eastman, "The Toughest Transistor Yet", *IEEE Spectrum*, May 2002, pp. 28–33.
93. K. Maruhashi, M. Ito, H. Kusamitsu, Y. Morishita, K. Ohata, "RF Performance of a 77 GHz Monolithic CPW Amplifier with Flip-Chip Interconnections", *IEEE MTT-S International Microwave Symposium Digest*, Baltimore, June 1998, pp. 1095-1098.
94. Endwave - http://www.endwave.com/Flip_Chip.html
95. HD Communications Corp. - http://www.hdcom.com/vbga.html

Index